Springer Series in Advanced Manufacturing

Series Editor

Professor D.T. Pham
Manufacturing Engineering Centre
Cardiff University
Queen's Building
Newport Road
Cardiff CF24 3AA
UK

Bernard Grabot • Anne Mayère • Isabelle Bazet
Editors

ERP Systems
and Organisational Change

A Socio-technical Insight

 Springer

Professor Bernard Grabot
Laboratoire Génie de Production (LGP)
Ecole Nationale d'Ingénieurs
  de Tarbes (ENIT)
47, Avenue d'Azereix
65016 Tarbes Cedex
France

Professor Anne Mayère
Laboratoire d'Etudes et de Recherche
  Appliquées en Sciences Sociales (LERASS)
114 route de Narbonne
31077 Toulouse Cedex 4
France

Associate Professor Isabelle Bazet
Centre d'Etude et de Recherche Travail,
  Organisation, Pouvoir (CERTOP)
Université de Toulouse-Le Mirail
Maison de la Recherche
5, allées Antonio Machado
31058 Toulouse Cedex 9
France

ISBN 978-1-84800-182-4   e-ISBN 978-1-84800-183-1

DOI 10.1007/978-1-84800-183-1

Springer Series in Advanced Manufacturing ISSN 1860-5168

British Library Cataloguing in Publication Data
ERP systems and organisational change : a socio-technical
  insight. - (Springer series in advanced manufacturing)
  1. Organizational change 2. Management information systems
  I. Grabot, Bernard II. Mayere, Anne III. Bazet, Isabelle
  658.4'06
ISBN-13: 9781848001824

Library of Congress Control Number: 2008925846

Cover design: eStudio Calamar S.L., Girona, Spain

Printed on acid-free paper

9 8 7 6 5 4 3 2 1

springer.com

Contents

1

The Mutual Influence of the Tool and the Organisation

Anne Mayère[1], Bernard Grabot[2], Isabelle Bazet[1]
[1]University of Toulouse 3 IUT A, LERASS
[2]University of Toulouse ENIT, LGP

1.1 Introduction

Enterprise Resource Planning (ERP) systems are now the backbone of the information systems in most large and medium companies, but also in many administrations. According to a study by ARC Advisory Group Inc., Dedham, Mass., the worldwide market for ERP systems is expected to grow at a 4.8% compound annual rate, rising from $16.7 billion in 2005 to more than $21 billion in 2010. This success is linked to several factors, and mainly to their expected ability to address the main limitations of former legacy systems – most of them interrelated – including coexistence of pieces of heterogeneous software, difficult evolution, lack of data and process integration, or high cost of maintenance. Moreover, the libraries of business processes included in the ERP packages are supposed to make possible the adoption of "best practices" allowing strong improvements of company performance.

In spite of these hopes, implementing an ERP is still considered as a risky and difficult project: many ERP implementations indeed suffered from excessive delays and costs but also from difficult appropriation by users, with consequences which could lead in extreme cases to a partial loss of control by the companies of their daily activities.

In the context of a peak of industrial demand due to the Y2K problem and to the introduction of the Euro in Europe, this situation was reported in a huge literature at the end of the 1990s, both from the technical and human science perspectives. After this period of effervescence, it became clear that, as a tool supporting most of the activity of the organisation, an ERP could not be considered only as an information system making human activity more efficient or as a technical artefact unable to cope with the social reality of organisations. In this new and perhaps more stable context, identifying the various dimensions of ERP systems and their impact both on the technical and social aspects of organisations becomes in our opinion a major and still open problem...

The aim of this book, written by engineers, computer scientists, consultants in organisation, sociologists, economists, and researchers from information and communication sciences, is to compare different views on ERP systems, taking into account the interaction between their technical interest, the constraints they set by their unavoidable interaction with the organisation and the individuals, but also their potentialities as tools to increase the understanding that individuals and groups can have of their organisations.

This interaction between ERP systems and organisations is particularly important because of the paradoxical abilities of these systems: on one hand, they can promote new processes allowing one to improve the daily activity of the company, while on the other hand they are supposed to be configured according to the specific requirements of each company. Such technical systems are therefore the result of an at least two-step design: design of the "envelope system" by the editor's teams first, with what can be analysed as an implicit or more explicit model of an "efficient organisation", notably through supposed "best practices"; finalisation of this design then, by the consultants and internal specialists, who configure the ERP software with the support of dedicated tools and methods. An ERP system is thus a "social construct" taking its definitive form all along this two-step process.

In order to explore the ambiguity of this relationship between the information system provided by the ERP and the organisation, we have chosen to explore three couple of notions which are for us at stake in this mutual adaptation process: integration versus communication, unification versus interpretation and alignment versus adaptability.

1.2 Integration Versus Communication

Integration is one of the major objectives of ERP systems, within the company, between functions or departments, but also outside the company, between business partners like customers, distributors, suppliers or sub-contractors.

But what is integration? Through definitions shared by management and computer sciences, it is often considered as creating a seamless flow of materials, finances, information and decisions in order to decrease waste due to multiple loose interfaces between islands of efficient activity processing. According to such definition, information system integration is closely related to the efficiency of the business processes inside and between firms. Two interrelated issues are linked to this overall evolution: models, as those promoted by Gilmour (1999), Cooper et al. (1997), SCOR (SCC, 2007) or EVALOG (2007) but also those included in the libraries of each ERP system, and tools allowing to make these models operational, mainly ERP systems and APS (Advanced Planning Systems) (Stadtler and Kilger, 2000).

On the technical side, integration is considered as a way to coordinate distinct entities working for a common goal, the organisation being often seen as a consequence of this coordination. This point of view, which has its origin in control science (see, for instance, Mesarovic et al., 1970; Le Moigne, 1974), considers that a hierarchical decisional structure allows one to cope with two problems or organisations: the mass of information to process, and the coordination of sub-entities. Figure 1.1.a schematises a purely centralised system. Different entities of the lowest level are coordinated at an upper level by another decision centre. Coordination is considered to be performed by mean of two information flows in opposite directions: top-down instructions aiming at making the behaviour of the lowest level decision centres consistent, and bottom-up follow-up information allowing one to inform the coordinator. In order to keep this information flow manageable, several coordinators are defined when numerous entities have to be coordinated, which requires then the definition of upper levels for "coordinating the coordinators". According to this framework, the different layers can be characterised by the horizon of their decision-making: long term for higher level, then middle term, short term and real time. At each level, degrees of freedom have to be preserved to allow a decision centre to react to unexpected events, within the boundaries defined by its coordinator.

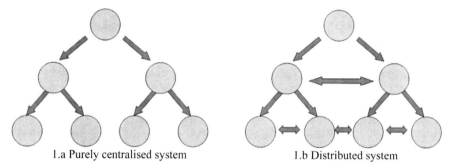

1.a Purely centralised system   1.b Distributed system

Figure 1.1. Communication in a decision system

Methods like GRAI (Doumeingts et al., 2000), CIMOSA (Vernadat, 1996) or GERAM (1999) have for instance used these concepts in the area of enterprise modelling. In GRAI for instance, each decision centre sends decision frames (mainly composed of objectives and means) to the lower level decision centres it coordinates. Reference models (see for instance Roboam, 1988) allow one to define a "consistent" way of performing activities, independently from the human actors present in each decision centre.

This general paradigm can be slightly modified in order to increase its efficiency. Figure 1.1.b shows for instance a structure where decision centres of the same level can adjust their decisions by mutual agreement. From the point of view of control theory, the two structures described in Figure 1.1 have complementary advantages: a purely centralised system is simpler, but each adjustment between decision centres of the same level requires to be performed by the coordination-higher level decision centre, which may induce delays. Communication between different decisional levels should decrease in the second case, but mutual adjustment may require arbitration difficult to obtain between decision centres of the same level.

The integration provided by ERP systems – both at the informational and process levels – is clearly consistent which such points of view: each entity of the structure can have access in real time to updated information, and each high level decision centre can have access to aggregated information from the immediately lower level, but can also "drill down" the information system in order to have access to more accurate information. Similarly the activity of each entity is framed by its role in pre-determined processes. In this respect, ERP systems are designed for piloting the organisation, and more specifically for answering the core issue of accurate reporting, in the contemporary context of the growing importance of financial profitability. As a consequence, it is most of the time considered that ERP systems reach the objective of integration through centralisation, since they speed up communication within a given level, but also between levels. On the other hand, a consequence of their use is that communication between entities of the same levels is no longer strictly necessary, the reaction delays between levels being decreased by the use of automated data processing. It is possible to say that ERP systems create a "temptation of centralisation" the consequence of which can of course be to decrease the decision flexibility of lower levels decision centres, and to increase the weight of centralised control.

Social sciences discuss such an analytical framework, and at least a part of the research community criticises what can be seen as a "machine like" model of the organisation. This criticism refers more specifically to three related arguments.

The first argument deals with the very definition of information. The data stored in the integrated database of an ERP do not make sense – that is to say, become "information" – through a "natural" or automated process. This requires notably what A. Giddens has called a "common sense understanding" or mutual knowledge, which involves "first, what any competent actor can be expected to know (believe) about the properties of competent actors, including both himself and others, and second, that the particular situation in which the actor is at a given time, and the other or others to whom an utterance is addressed, together comprise examples of a specific type of circumstance, to which the attribution of definite forms of competence is therefore appropriate" (Giddens, 1986, p 89). The "communication in a decision system" model does not address the question of sense making. This is, however, an important issue concerning integrated information systems supposed to be used by a great variety of professional groups, having different background, knowledge and competences, dealing with different activities, and all producing relevant information out of shared databases.

Another implicit hypothesis is that the information required at a global level (namely, the management level of the firm) presents basically the same characteristics as the information required at a local, more activity-oriented level. In other terms, information that can be more or less detailed is yet supposed to share the same structure. However, research results (see Bazet and Mayère Chapter 4) show that information required at the operational level, which is often very contextualised, can be different from what is required to pilot the organisation. So, the question is not only where the decision centres are, but also what is the relevant information for them?

Thirdly, such models are also based on a fairly strong hypothesis concerning the rationality of firm decision. James March and other researchers have underlined the role of ambiguity and vagueness in decision processes, particularly in organisations characterised by complexity and uncertainty, which is often the case for contemporary firms (March, 1991, Mayère and Vacher, 2005). The related question is to know whether ERP systems do offer the opportunity to deal with uncertainty, adaptability and possible contradictory logics within a firm and between firms involved in the same supply chain.

Coming back to the control theory model, experience shows that reality is less simple since most ERP systems provide both the possibility of centralisation and the facilities allowing one to manage a distributed system, for instance with workflow or groupware tools. Moreover, it is clear that process orientation is more efficiently implemented in a distributed system. In this respect, centralisation could be a consequence of implementation choices rather than of the intrinsic design of ERP packages.

In the control science paradigm, integration should allow improved decisions to be made on the base of the available information in order to control the material and finances flows. This view, also coming from early applications of system science to organisations (Le Soigné, 1975), is schematised by the left layer of Figure 1.1. Within this paradigm, the "enterprise system" can be sub-divided in several sub-systems, namely decisional, physical and informational sub-systems, to which we would add the "financial" sub-system. This view, which can be found with slight differences in enterprise modelling methods like GRAI or CIMOSA, has for a long time been considered as sufficient for most engineers. Very consistent with the general frameworks of Figure 1.1, it provides a sufficient mapping for making a clear distinction between improvements regarding layout or material handling (physical system), information processing or software (informational system) or decision making (decisional system).

In parallel, researchers in human science have questioned the links between individuals and organisation, analysing the relations between these organisations (by functions, projects, matrix structures, etc.) and the communication and social links between individuals (which could be considered as part of right hand layer in Figure 1.2.).

The enterprise has multiple dimensions, all related, which can as a consequence hardly be considered individually. Having a technical view on decision or information, bases of communication, without considering their implications at the organisational and at the human level is impossible. Similarly, focusing on the interactions between individuals or between individuals and software without any reference to the technical background of these exchanges is vain. The interaction between the two extreme layers can be considered as being provided by the process layer which defines the activities to be performed with reference to the required data, resources and actors or groups of the organisation. The integration provided by process-oriented models, supported in practice by ERP systems, is for instance clear when considering case-tools for Business Modelling like ARIS (Scheer, 2000), which suggest a business process model that gathers views on data, functions, resources, organisations and added value. Nevertheless, these models only show which entities of these various views are involved in a given process.

Such a representation may be an interesting attempt to take into account the different dimensions concerned with the transformations at hand. However, social science researchers underline the necessity for questioning deeper, and in a more systemic way, the interaction between the organisation, its production process, and the information sub-systems, none of them being independent of the others. More precisely, this would require specifying what is meant by the "organisation" level. This would also require taking into account the design process of the various sub-systems, which has to be thought of as a social process. What are their mutual influences, how to develop the most efficient process regarding a given organisation, or contrarily what is the relevant organisation for performing a process, are questions that the enterprise has to answer daily during its perpetual improvement activity. Such questions clearly address socio-technical issues.

After many years during which only the positive aspects of integration were emphasised, especially in the engineering field (and perhaps as a consequence of these deficiencies), integration is now considered a more bivalent paradigm, which may induce new problems. This is especially clear in Supply Chain Management, where the autonomy of the partners together with their need for confidentiality on strategic information set problems for using highly integrated tools like APS (Advanced Planning Systems - APS (Stadtler et al., 2000)). Nevertheless, as shown by the research on human science, perverse effects were also present inside the companies, integration being the cause of changes in the communication protocol between partners, which were not always expected or desirable.

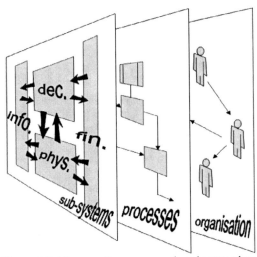

Figure 1.2. Views on the company: three layers at hand

Chapters 2 and 3, by Jonas Hedman/Stefan Henningson and Carole Groleau, illustrate, at the individual then organisational level, that integration may be in some cases an inadequate substitute to communication, considered as the real challenge of organisations. Anne Mayere and Isabelle Bazet show in the third chapter that integration may lead to increased control of individual works through standardisation of the confrontation spaces, leading often to local inefficiencies.

We will explore further these contradictory dynamics through the second couple of notions at stake: unification and interpretation.

1.3 Unification Versus Interpretation

Allowing several entities to work together to reach a common goal (these entities being either different departments of a company or different companies in a supply chain) requires adequate communication means. In the technical field, making communication possible and efficient has for a long time been interpreted in terms of software solutions for an efficient information exchange. Indeed, each entity (department or company) based its daily work on the use of pieces of software which were in the past often communicated poorly one with another. Costly and hardly maintainable interfaces, together with communication protocols, were first developed to cope with this problem. EAI (Enterprise Application Integration) systems provided more flexible and productive tools to allow the development of interfaces between different pieces of software. It became then clear that exchanging data is a required but not sufficient condition for communicating, the main problem being to give sense to the available information, i.e. to have a common interpretation. A more comprehensive concept of interoperability then emerged, based on the definition of communication layers (at least, technical, syntactical and semantical). Great effort was consequently directed towards ontologies (Gruber, 1993), aiming at supporting the emergence of a common – unified – interpretation of the information, either exchanged by people or pieces of software. Unification of the sense is nowadays usually considered as one of the major conditions of interoperability.

ERP systems, based on a unique database and on business processes which integrate the various functions of the company, were supposed to facilitate communication through information sharing. Nevertheless, allowing different users or group of users, possibly belonging to different companies, to have direct access to the same information, or to share the same processes is not enough. Social scientists would say that the question is: do the actors share a common reference framework, so as to make relevant information from the recorded data? On the technical side, this requires unification of the interpretation, but is there sufficient evidence to make sure that such unification is possible and desirable? The results of a research program with indepth investigations in an international group show, for instance, that the purchase process is possibly different, according to the market structure, between European and North-American plants; when a similar computerised process has been set up, new constraints appear with possible loss of efficiency at a regional level (see Chapter 4 by Bazet and Mayère). Whether unification can – and should – really pass cultural boundaries is a subject which is nowadays more and more often addressed in the field of ERP systems, "culture" being possibly linked to companies, technical backgrounds or countries (concerning this issue, see Chapter 11 by Motwani et al.).

Considering this evolution, one may say that computer and engineering scientists and social scientists tend to converge on a similar fundamental question, which is: how to make sense from data to information. However, from the social science point of view, this question is not only an individual but a collective and organisational one: sense making in action is the product as well as the condition of cooperation. So, the question is not only to know whether the data recording and processing are coherent with the mutual knowledge and the action on hand, which can differ according to professional activities and location. The question deals also with the design of communication processes and their compatibility with organisational functioning, which is basically of social nature, with "global" characteristics in relation with Society structures, combined with "local" features.

Several works suggest now that interpretation, usually considered as a source of incomprehension by engineers, may provide necessary degrees of freedom allowing adapting a rigid structure to an ever-changing and multi-faced reality. In that sense, interpretation cannot be given to or imposed on several heterogeneous groups through unification, for instance by the design of a common ontology. Interpretation is built by a social group in a given context, in consideration with the available knowledge.

The second part of this book intends to illustrate that even in the presence of ERP systems, interpretation remains one of the main mechanisms of social interaction. James Taylor suggests considering both the ERP and the organisation as texts subject to interpretation, whereas Ben Light shows that some of the problems during ERP implementation and use can be explained by different interpretations of the tool by their designers and users. Séverine le Loarne and Audrey Becuwe illustrate through a real case that ERP systems can be the basic tools of legitimisation processes based on an interpretation of their use, whereas François Marcotte shows that paradoxically, the large amount of data available in ERP systems may result in a temptation to opportunistically redefine informal processes, especially in uncertain environments.

1.4 Alignment Versus Adaptability

As stated previously, ERP systems provide data and processing integration, which can be seen as an opportunity for unification. Similarly, ERP systems are based on business processes, and may therefore allow alignment, i.e. standardisation of practices according to pre-defined standards. Indeed, the alignment of the business processes of a company on "best practices" included in the ERP was considered as one of their basic interests some years ago, whereas customisation and adaptation of these systems were seen as a major cause of difficulties during the implementation phase.

According to an engineering point of view, the concept of "best practices" is justified by the idea that a large part of the activity of industrial companies can be efficiently performed using invariant processes. This point of view is comforted by a strong current of thinking coming from continuous improvement methods, which has had great success these last 10 years. Methods like just-in-time, lean manufacturing, 5S, 6 sigma, competence management, knowledge engineering, etc. have been successfully used in numerous companies, which has perhaps reinforced the idea, already present in engineering, that improvement comes more easily from adoption of universal, validated techniques than from the emergence of specific approaches dedicated to a given company. Nevertheless, such dedicated techniques are potentially more compliant with the know-how and culture of a given company, and may become a competitive advantage since by definition, they are not shared by competitors.

Implementing standard processes was considered as a strong point of ERP systems, even if the difficulty of this task has been underlined by many authors for a long time. For many years the justification of this adoption of "external" processes was only questioned in a few works. This is certainly not so clear now, and even in the field of business processes or ERP systems implementation, several authors underline that best practices may fail to gather the knowledge-based specificity of a company, that makes its success. Moreover, the cultural and social issues of the alignment on best practices seem to suggest that the worldwide industrial culture does not supplant local realities in all the countries. Finally, there could be external costs linked to supposed-to-be best practices that could soon give rise to renewed debate (see for instance the just-in-time paradigm with respect to environmental issues).

This statement has been made much earlier in the Human Science domain, more sensible to the problems of adoption and appropriation of new techniques than the Engineering side. Moreover, social sciences have shown that such an adoption of supposed universal techniques, which are basically organisational and therefore social, requires important organisational work (de Terssac, 2002). That is to say, the organisation itself is the object of reflexivity. With such methods, employees are asked to contribute to the organisation re-design, and to evolve in relation with this change. Such organisational work is very demanding both at an individual and at a collective level; it may ask for a new conception of what is part of the job, the competence required, and the necessary cooperation scope.

Based on real cases, Bernard Grabot will show how business alignment can be a source of improvement but also how some authors consider that such a standardisation process cannot provide a competitive advantage, possibly brought by customisation. Valérie Botta-Genoulaz and Pierre-Alain Millet illustrate the link between business alignment and maturity of the company regarding Information Technology tools, whereas Jaideep Motwani, Asli Akbulut, Thomas V. Schwarz and Maria Argyropoulou show, through a comparison between ERP implementations in USA and Greece, that cultural factors should be taken into account during business process selection.

1.5 New Developments in ERP Integration

The paradigm linked to the implementation of ERP systems has evolved over fifteen years. In the 1990s, ERP systems were considered as a major opportunity for performance improvement, thanks to data and activity integration and standardisation, but also as a source of increased control of heterogeneous companies, which were opportunistically grouped into holdings. At the same time, many factors caused a high ratio of unsatisfactory implementations, up to the point that ERP systems were (and are still?) largely considered as monolithic tools for standardisation, crushing the specificities of companies and the motivation of individuals.

Knowledge of these systems, especially their interests and limitations, has increased with experience and they are now considered in a more balanced way than some years ago: ERP systems are large and complex systems, which deeply modify the activities and organisation of the companies in which they are implemented. Their collision with organisation and individuals can be an opportunity for improving existing processes and behaviours. They can also be a way to identify discrepancies between real practices and standard processes which can be a source of motivated customisation. Such adaptation of the system, still the object of suspicion, is nevertheless more and more considered as possible and necessary. Configuration and interoperability with external systems may allow one to cope with the well known problems that come from specific developments. It is certainly this balance between standardisation and adaptation that will be the major challenge of the next generation of ERP systems.

As a conclusion, a working track could be that ERP systems have to become social systems able to address both their technical and organisational challenges, questioning the compatibility and possible synergy between information system efficiency and organisational work, taking into account the information and communication issues, and the technology as a social construct (Feenberg, 2004). This means questioning at each stage of its design the implicit hypothesis regarding the social structure, the organisation of labour, the social relations and related communication activities, and the sense making process. Such an approach should require "opening the black box", to include thinking of the users, both final and intermediary, as agents of the technology design in a renewed sense compared with what is often the case currently. This will ask for combining and developing technical and organisational knowledge through a renewed collaboration between computer and engineer scientists and social scientists, that is to say, going further in the debate between different disciplines. This book is an attempt at such a process, which still has to go further.

Acknowledgement: The authors would like to thank the French CNRS, and more specifically the "Information Society" Research Program, as well as the "Maison des Sciences de l'Homme et de la Société de Toulouse" and the I*PROMS Network of Excellence for their financial support for organising the seminars that resulted in this book.

1.6 References

Cooper M, Lambert D, Pagh J, (1997) Supply Chain Management: more than a new name for logistics. International Journal of Logistics Management, 8(1):1–14

Dougmeingts G, Ducq Y, Vallespir B, Kleinhans S, (2000) Production management and enterprise modelling. Computers in Industry, 42(2–3):245–263

EVALOG, (2007) Global EVALOG frame of reference, http://www.galia.com

Feenberg A, (2004) (Re)penser la technique. Ed. La Découverte, Mauss

Geram, (1999) GERAM: Generalised Enterprise Reference Architecture and Methodology, version 1.6.1, IFIP-IFAC Task Force on Architecture for Enterprise Integration, March

Giddens A., (1994) Les conséquences de la modernité. Paris: L'Harmattan

Giddens A., (1986) New rules of sociological methods: a positive critique of interpretative sociologies. Hutchinson & Co. Publishers, London

Gilmour P, (1999) A strategic audit framework to improve supply chains performance. Journal of Business and Industrial Marketing, 5(4):283–290

Gruber T, (1993) A Translation Approach to Portable Ontology Specifications. Knowledge Acquisition, 5(2): 199–220

Le Moigne JL, (1974) Les systèmes de décision dans les organisations. Presses Universitaires de France

March JG, (1991). Décision et organisation, Dunod

Mayère A, Vacher B, (2005). Le slack, la litote et le sacré, Revue Française de Gestion, hors série "Dialogues avec James March": 63 – 86

Mesarovic MD, Macko D, Takahara T, (1970) Theory of hierarchical, multilevel systems, Academy Press

Poole MS, DeSanctis G, (1990) Understanding the use of group decision support systems: the theory of adaptative structuration. In Steinfield et Fulk (Ed.), Theoretical perspectives on organisation and new information technologies, Ed. Sage: 173–193

Roboam M, (1988) Modèles de référence et Intégration des méthodes d'analyse pour la conception des systèmes de production. PhD thesis, University of Bordeaux I

de Terssac G, (2002) Le travail: une aventure collective. Toulouse: Octarès Editions

Scheer AW, (2000) ARIS - Business Process Modelling. Third Edition, Springer

SCC, (2007) Supply Chain operations reference model: overview of SCOR version 7.0. Supply Chain Council, http://www.supplychain.org

Simondon G, (2005) L'invention dans les techniques. Cours et conférences, Seuil

Stadtler H, Kilger C, (Eds.), (2000) Supply Chain Management and Advanced Planning. Springer.

Vernadat F, (1996) Enterprise modelling and integration, principles and applications. Chapman & Hall, London

ERP Systems in the Extended Value Chain of the Food Industry

Jonas Hedman[1], Stefan Henningson[2]
[1]CAICT, Copenhagen Business School, and University College of Borås
[2]CAICT, Copenhagen Business School

2.1 Introduction

Enterprise Resource Planning (ERP) systems are one of the most important developments in corporate information systems (Davenport, 1998; Hitt et al. 2002; Upton and McAfee 2000) and in Information Infrastructure (II) (Hanseth and Braa 2001) during the last decade. The business interest in ERP systems can be explained by the benefits associated with the implementation and utilisation of ERP systems (Robey et al., 2002). The benefits are related only in part to the technology, most of these stemming from organisational changes such as new business processes, organisational structure, work procedures, the integration of administrative and operative activities, and the global standardisation of work practices leading to organisational improvements, which the technology supports (Hedman and Borell, 2003).

The implementation of ERP systems is a difficult and costly organisational experiment (Robey et al., 2002). Davenport (1998) described the implementation of ERP systems as "perhaps the world's largest experiment in business change" and for most organisations "the largest change project in cost and time that they have undertaken in their history". The costs and time frame related to implementing an ERP system can be illustrated by the case of Nestlé, which had invested, by the end of 2003, US$ 500 million in an ERP system. In 1997, the American subsidiary started the project and in 2000 the global parent decided to extend the project into a global solution (Worthen, 2002).

One of the goals many companies strived for was homogenous and standardised corporate information system and II. With the result in hand, we can see that the foreseen architecture never was accomplished. Rather complex II emerged as a result of pressure and changes in, to the organisation, external and internal context (Hanseth and Braa, 2001). Nevertheless, the business needs that originally laid the foundation for approaching consolidated large scale systems still prevail. Managers are still seeking organisational transparency, customers are still demanding one global partner, production and logistics may still be smoother with appropriate coordination, and in addition information and process inconsistencies are lurking across the enterprise. As a result there has been an increased focus on ways to make possible coordination and cooperation between business units, customers, and suppliers. The advantages of integrating the extended value chain are apparent in most industries. By integrating business processes from the end consumer to original suppliers in terms of products, services and information the participating organisation can provide additional value for the consumer and/or supplier and thereby increase the value of the entire value chain. Forward and backward integration enables that higher efficiency and effectiveness in areas such as scheduling, transactions and planning can be leveraged (Lubatkin, 1988). Still, the reality tells us that in many cases well managed contemporary companies are not fully integrated with their descendants and antecedents in the value chain.

It is impossible to achieve an effective supply chain without information and communication technology (ICT) (Gunasekaran and Ngai, 2004). The beneficial effects are heavily dependent on the ability to integrate information systems (IS) appropriately (Henningsson, 2007). The development in ICT during recent decades, such as ERP systems, has set the ground for global integration initiatives as it is now possible to create the II that are necessary in geographically spread value chains. ICT is the base to create the integrated extended value chain (Gunasekaran and Ngai, 2004). In this sense, ERP systems implementation and IS integration reflects the strategic decisions regarding integration in the extended value chain and integration initiatives reflect the ambition to integrate with peers.

Much is written on ERP systems, intra-organisational value chains, and internal integration and how to make it work efficiently. Less is known about the extended value chain with focus on the whole chain from initial producers to end consumers (Browne et al., 1995; Jagdev and Browne, 1998) or the role of ERP systems in the context of extended value. Methodologies like Business Process Re-engineering does not apply well to external business processes as different corporations often operate autonomously: there is no higher authority to orchestrate a top-down approach. In this article we present a study of four product flows (milk, pork, sugar, and peas) involving nine organisations active in the extended value chain of the Swedish food industry. Our purpose is to describe how they integrate and use their ERP systems and on which foundation they assumes their strategies. In doing so, we will provide further insight to the understanding of ERP systems in the external integration in the extended value chain. In the next section we present how the study was carried out before turning to the theoretical framework in which our research contribution should be fitted. Further, we present the findings and our theoretically grounded analysis that aims to clarify key criteria for choosing to or not to integrate corporate II in the extended value chain.

2.2 Research Methodology

The empirical phase of the work aimed at collecting data in relation to the theoretical framework. Nine case studies were performed along the extended food chain, including three farmers, four food producers, one corporate function of one grocery chain, and one large grocery store. The research questions were organised in two parts. The first part was loosely centred upon customers, products, business processes, work activities, organisational structure, and suppliers in order to gain background on each company. The second part addressed the use of ERP systems and integration of IS between customers and supplier. The main method used is interviewing. In total 13 semi-structured interviews were made. Interviewees were selected in order to provide a broad representation of those involved. In most cases the interviews were made by two master students which lasted between 20 and 100 minutes. Based on the data material nine case stories were written up. The cases are:

- Askliden AB is a milk producer, with 250 milk cows.
- Bramstorp Gård AB produces sugar beets and peas.
- Coop Norden is the corporate function of the second largest grocery chain.
- Danisco Sugar's facility at Örtofta refines sugar beets into raw sugar – has a monopoly.
- Findus AB is specialised in frozen food, such as vegetables (illustrated by peas), meat and fish.
- ICA Tuna is a local grocery store and belongs to the ICA group.
- Skånemejerier is a cooperative owned and a leading actor among dairy products.
- Swedish Meats is the leading slaughter house in Sweden and also a cooperative.
- Tygelsjö Mölla is a pig farmer, who delivers 4500 piglets to Swedish meat.

The cases were selected based on the four product flows: milk, pork, sugar, and peas. The cases are presented along the four product flow. The product flows were in turn chosen based on some unique features regarding need for integration along the value chain, e.g. planning horizon and harvesting are critical time constraints for pea farming. A number of actors and products are not included, for example end consumers and governmental agencies, such as the Swedish Health Department and EU.

We used Yin's (2003) pattern-matching analysis method, whereby the empirical observations were "matched" and compared with theoretical concepts. Being a case study aimed at generalising towards theory (rather than population), we used the empirical findings to "challenge" existing theory and concepts related to ERP systems use and integration in the extended food chain. The phenomenon under investigation is integration mediated through II and ERP systems along the extended value chain. Thus, we will not make any claims regarding the individual cases or the products – only towards the integration or lack of integration of the food chain.

2.3 Integrating the Extended Value Chain with ICT

ICT has gone through some dramatic development during recent decades. The development enables companies to work and structure their business processes in new ways (Jagdev and Browne, 1998). However, integrated companies are dependent on complex integrated IIs, such as ERP systems, CRM systems, and SCM systems.

2.3.1 Integrated Information Infrastructures

Only a decade ago many companies strived for homogenous and standardised ERP systems that should be the informational backbone of the corporation and seamlessly integrate business processes and information flows throughout the whole supply chain. With the result in hand, we can see that in spite of the substantial efforts put into the quest, the foreseen architecture never was realised. Instead rather complex IIs emerged as a result of both technical and organisational issues (Hanseth and Braa, 1998). Despite the unsuccessfulness of the great enterprise-wide IS in the effort to consolidate corporate IS constituents into one large scale system, the business needs that originally laid the foundation for approaching these systems still prevail. By various approaches and techniques, organisations search to integrate not only their internal processes, but also processes that take part of an extended value chain.

The II of today's companies consist of a growing pile of systems that specifically target various aspects of the business, including Customer Relationship Management (CRM), Enterprise Resource Planning (ERP), Supply Chain Management (SCM), Business Intelligence (BI), Content Management (CM), Portals, Computer-aided Design (CAD), Embedded Systems, and Network and Collaborative systems. The terms of these systems tend to vary as vendors and consultants launch new marketing efforts, and trends come and go. However, the business needs they address tend to show more stability over time.

2.3.2 Information Infrastructures in the Extended Value Chain

Management and coordination of the internal value chain is a well researched topic (Konsynski, 1993). Less is known, however, about the strategy of corporations taking part of a larger chain that spans cross-organisational boundaries, the extended value chain (Markus, 2000). The development in ICT, in combination with extended pressure of globalisation, environmental consideration and transformation of organisational structures, is said to transform organisational boundaries, blurring the frontiers to customers and suppliers (Browne and Jiangang, 1999; Markus, 2000). As the technological platform continues to develop, the foundation for extending the corporation's business process becomes more and more relevant. Manufacturing companies can no longer be seen as individual systems, but rather as participators in an extended value chain (Browne and Jiangang, 1999). Optimising this value chain is one major challenge in order to achieve business success (Jähn et al., 2006). Establishing the appropriate II has been found closely related to many of the potential benefits that can be obtained by combining organisational units (Henningsson, 2007). Also in the leveraging of potential benefits from integration into an extended value chain creating the integrated II is considered a key issue (Gunasekaran and Ngai, 2004). The task has been found cumbersome, as corporations suffer from not having sufficient knowledge on what type of infrastructure is required to achieve the desired supply chain (Gunasekaran and Ngai, 2004).

In order to set the ground for the extended value chain, somehow the II of the organisations have to be integrated. "Surprisingly, very little literature directly defines integration" (Schweiger and Goulet, 2000: 63). Creating an integrated information infrastructure roughly denotes the creation of linkage between previously separated IS (Markus, 2000) at technical, business process, and business practice level (Konsynski, 1993). Although IS integration normally is thought of in the context of modern, global corporations doing real-time business with its partners, the idea of IIs is not new. The idea emerged during the 1940s to 1960s and serious discussions on how to replace existing islands of isolated systems with new, totally integrated systems may be traced back to the 1970s. Corporations seek business integration basically because customers and suppliers demand and expect it. For intra-organisational integration the key benefits are often referred to such things as providing the customer with one single interface and harmonise production and logistics throughout the corporation (Markus, 2000). The drivers for external integration are somewhat different. Business drivers for intra-organisational integration includes higher ability for organisational learning, better ability to respond to market change, and in the end more efficient management due to new or smoother information flows (Konsynski, 1993).

Integration of the extended value chain addresses both internal and external integration affected by different forces and have different goal. The bottom line is efficiency and effectiveness improvements in order to gain competitive advantage.

2.4 The Food Industry as Extended Value Chain

Our empirical data stems from a study of nine companies involved in the Swedish food industry.

2.4.1 The Swedish Food Industry

The number of end consumers is just above ten million consumers, whereof one million Norwegians, who cross the border to Sweden to shop for food and alcohol. The Swedish market is dominated by three large retail chains, ICA, Coop, and Axfood, with a total market share of 72%. ICA and Axfood are privately owned, whereas Coop is a cooperative owned by the consumers. During the last few years two new low price retail chains (Lidl and Netto) have entered the market. They are mainly taking market share from Coop.

There are several major food producers in the southern part of Sweden – Skåne, such as Procordia Food AB, Findus Sverige AB, Skånemejerier, and Pågen AB. Skåne is the most important agricultural area of Sweden with some 8700 farmers.

The main food products from Skåne are different types of crops, dairy products, rape, sugar beats, and meat.

In addition to the companies directly involved in the extended value chain, there are several other important actors in the food industry. These actors are not actively involved in production, but have the potential to influence the end customers and their preferences, the products produced by the farmers or a general influence in the whole chain. Other actors are: KRAV (key player in the organic market), European Union (EU) and its Common Agricultural Policy (CAP), National Food Administration, Consumers in Sweden, and Agricultural Universities.

In the following sections, we present empirical data from the nine cases along the four product flows. The focus is on various types of II integration among the actors. The specific information flow and ERP systems of the four products (milk, meat, sugar and peas) from the farmer, through the producer and to the retailer, will also be described.

2.4.2 Milk Flow

Milk production at Askliden AB is supported by milk robots and automatic feeding machines, almost without human interaction. The data collected by the milk robots (amount and quality, etc.) is linked through an advanced IS to the Swedish Dairy Association (SDA), who analyses data and provides feedback, e.g. what to feed each individual cow. SDA and Askliden also diffuse data and information to other milk producers, such as quality of milk. In addition, Askliden uses a number of other IS to support their business. For instance, PC-Stall Journal to manage all their livestock and Genvägen which used to pick out the best bull for each cow. The data is not passed to the Skånemejerier.

The entire production is sold to Skånemejerier. The price is based on quality (fat and taste). When the milk arrives from the farmers (about 900 dairy farmers) at the dairy in special halls it is checked, and pumped into storage silos. Thereafter it is processed. The milk is cooled down and the cream is separated from the milk. Both products are pasteurised and mixed together again to meet specific percentages of fat. Before the milk is packed it is homogenised. To support this Skånemejerier use Movex (a large ERP-system from Lawson) to handle logistics, purchasing, resource management, financial assets, maintenance, supply chain management and data warehousing. A system called EDI/Link-XLM is used to manage the electronic information flow (order, invoicing, and payment) to and from farmers and customers. It is fully integrated with Movex. In the beginning of 2005 the EU passed a law that required the possibility to track the origins of food products. To comply with this law, Skånemejerier implemented a system that could be used for extracting up-to-date packing data. When managing customers, Skånemejerier uses a CRM. The packaged milk is sold to the retail chains in the southern part of Sweden. In total, Skånemejerier have 1 million end consumers in Skåne.

2.4.3 Pork Flow

Tygelsjö Mölla specialises in pig breeding and has one costumer, namely Swedish Meat. The farmer makes an agreement once a year on production quotas. The production quota is 4500 Piggham pigs delivered on regular bases. The quality of the pig is based on percentage of fat in the meat. Low percentage of fat increases the value, because it makes slaughter easier. However, low fat percentage affects the taste in a negative sense. In order to benchmark the individual farmer Swedish Meats provide the farmers with PIGWIN. Tygelsjö Mölla use PIGWIN to compare their own productivity with other breeders. They also use a web portal supplied by Swedish meats with information such as the quality of the animals they have delivered, and how much Swedish Meats are willing to pay for these. In addition Tygelsjö Mölla informs Swedish Meat about how many animals they require for slaughter.

Swedish Meats is one of the largest slaughter houses in Sweden. It is also owned by the breeders. The information flow starts with communication between the farmer and Swedish Meats. The farmer notifies Swedish Meats via the Internet, SMS or telephone, on how many and what kind of animal that he/she wants to deliver. Swedish Meats uses several different systems to collect data about the animals, for example their weight, age and origin. All of the information from these systems is sent to their ERP system. Swedish Meats uses four at five systems when interacting with the farmers for handling payment, butchering notifications and so on. They also use a CRM (Contact Relationship Management) system when collecting information from the farmers which is used to keep track of all of 17 000 breeders. Swedish Meats collaborates closely with their customers concerning quality and relevant production information. Swedish Meats has decreased their client list from over 10 000 customers when almost every store was their customer, down to three customers (ICA, COOP and Dagab) and 100 industrial customers. Even though the system handles the whole process from the farmer until the meat is packaged and delivered, no detailed information is passed on to the customer. It is possible to have a continuous information flow from the origin to the end customer, if requested.

2.4.4 Sugar Flow

To grow and harvest sugar beet, Bramstorp Gård uses a web portal (www.sockerbetor.nu) supplied by Danisco Sugar. All information exchange between the farmer and Danisco is done through the portal. The information consists of, for example, invoices and dates for seed distribution. The information flow is more or less one-way, from Danisco to the farmer. Danisco, as the leading sugar producer in the region, uses SAP's ERP-system R/3 to cover the IT needs of the entire organisation, internally as well as externally. As we are focusing our study on the information flow concerning sugar beet, we will not discuss the company's internal systems (for maintenance for example), but instead concentrate on the part of the ERP-system that handles external information exchange. The SAP modules used in the sugar beet information flow are: Agri, Sales & Distribution, and Logistics. The Sales & Distribution module (SD) is used to handle the information exchange between Danisco and their customers, while the Logistics module aids the transportation of the processed product (feed and sugar) stored at Örtofta. Agri is used to control the delivery of beet from farmers by creating delivery plans. The module is connected to the web portal www.sockerbetor.nu. As in the Findus case, Danisco aims to guide the farmer on how to best cultivate sugar beet by providing information, for example appropriate PH levels, protecting against erosion, balanced fertilisation and numerous hints and tips on how to protect and salvage parasite infected crops and soil. After the sugar beet has been harvested and transported to Danisco's processing plants the sugar is extracted from the beet and mixed, thus breaking the information chain.

2.4.5 Pea Flow

When we investigated the Bramstorp case we found that when growing peas this is not controlled by the farmer. The production of peas is a very controlled and regulated process and dominated by Findus with a market share of 60%. The process has an 18-month time horizon, i.e. the foundation that is laid in March should produce a harvest in August the following year. To support this Findus has developed a concept called LISA (Low Input Sustainable Agriculture) which aims to structure the process and minimise the weaknesses. LISA is based on the selection of fields for growing peas by analysing the soil in different fields, picking the most suitable fields and monitoring the development of the crops while looking for signs of harmful organisms. The subsequent harvest and processing of the peas is also a highly controlled and automated process. It is Findus who controls the information gathering, and they more or less tell the farmer what and where to grow peas. Findus uses ERP-systems from both SAP (financials and administration) and Movex (logistics and production). They supply the farmers with information about which fields are suitable for pea cultivation, when to plant seeds, how much and what kind of fertilising. This information is extracted from Findus's databases, which are based on soil samples from the farmers' fields. This means that in many cases Findus knows more about a field than the farmer that owns it. In addition Findus even harvest the peas with their own machines. In the production at Findus's plant data about peas is gathered, such as quality and origin, so that Findus can provide feedback to the farmer. In the future even the end consumer can take part of this information, i.e. know from which field the peas have been grown. Today the information flow is broken when the peas are packaged for consumers. There is no integration between the Findus production system and the packaging system. Findus also collects a lot of information from the market and competitors, but this is not an issue for the pea information flow.

2.4.6 Retailers and Grocery Chains Information Flow

ICA Tuna is a private owned grocery store and a member of ICA. The individual store is (livegen) concerning the assortment. It is decided by ICA centrally. The individual grocery store does not manage any IS by themselves. The central IT group at ICA develop and manage all IS and IT support. The main system is an intranet called "Slingan". ICA Tuna is connected to it as well as all other ICA stores in Sweden. In "Slingan" they can access ICA´s central storage to make orders, browse products available, check prices, and communicate with other ICA stores and exchange ideas and tips. However, it does not work as they expected due to technical problems. For example, if working with an order and the connection breaks down the system does not cache the data and all the input has to be entered again. So, not all units use it and the most common way to place orders is by telephone. All payments and financial processes are done through a web based system. When ordering from ICAs central storage, the payment is automatically withdrawn from ICA Tuna's account.

The retail chains, i.e. Coop and ICA, have most of the power and control, since the have contracts regulating what each store can sell and does most of the purchasing. In addition to that they also control all information flows through central IS. The corporate chains decide on what products to have in each store. They try to control the food producers (Findus, Danisco, Skånemejerier, and Swedish Meat) by keeping them in a state of uncertainty by not integrating their information systems. The only information flow is related to order, invoice and payment. Not stock quotes. The upward integration with end consumers is not well managed. Coop collect data on consumer behaviour through their loyalty cards (MedMera). This has made it possible to direct offerings and marketing to the right customers. Coop uses an IT system to register articles that have been sold in a store each day. This information is stored in a Data Warehouse and is used as decision support for the purchasing department. The system makes it possible to see how much of a product is being sold and the effect of marketing campaigns. There is an IT support system for sales planning that can simulate additional sales. The system can provide information about how much Coop has to sell to lower the margin on a product and plan the product flow. This is especially important during a campaign when the number of products to be sold increases since a product takes time to get to the store. The system calculates how large a quantity the store has to buy before a campaign, what is needed and how much. In addition a number of computer systems contain and handle information about the products, where they are located, how much is in stock, pricing and so on. The price on merchandise is set early on in the process (after the product has been delivered from the supplier) and stays with the product the whole way to the store where the price information goes directly into the cashier system. The process of handling all this information about a product is very costly (about 1% of the product price). The system supports orders and placement of the product during delivery, and the financial flow that is connected to the product during the delivery from the purchaser to the store. There is a new system in Malmö where the personnel that drive the trucks in a big warehouse use headphones to get information of what to do. After the task is done they indicate this by entering the information directly in the computer system in the truck.

2.4.7 Summary

The Swedish food industry is to a large extent characterised by a number of oligopolies among the dairy chains, monopolies among the food producers, rigorous governmental regulations, and loyal end customers. However the competition has increased over recent years, due to new retail chains, deregulation of the global food market, and increased price sensitivity among end customers. Integration takes place mainly between the farmers and food producers. The key aspect for this integration is how sensitive the product is; cf. sugar beets and peas. Between the food producers and the grocery chains there is little integration. The only information given to the food producers is order quantity. They have no or little information regarding sales and stock, which hampers their production planning. At the end of the food chain huge amounts of data is collected through the loyalty cards.

Figure 2.1 shows the four product flows and the cases along the value chain. It also shows the integrated information infrastructure and where it is not integrated. Note that the integration flow converge at the grocery level, indicating that information flow concerning the products is handled in the same way here.

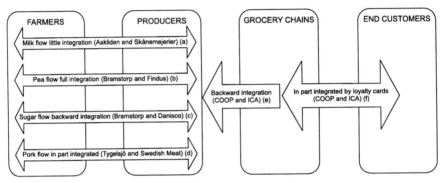

Figure 2.1. The Integrated Food Chain

2.5 Key Criteria for Integration Choices

With the aid of various ERP systems, detailed information about food products can be collected and distributed throughout the food chain. It is the producers and the retailers who dominate the information flow in the food chain, in large part because of their size. In most cases they control and provide the various systems that are used for data gathering and distribution. Retailers are dominant in terms of controlling the main retail outlets of food products in Sweden and the information concerning the products they sell. They try to influence the producers and the packing industry by pushing for the use of RFID (Radio Frequency Identification) tags on their products as well as added barcode information (EAN 128).The producers aim to control the production process to a large extent because they are responsible for the safety and quality of the food products by law, and at the same time want to maximise the production output of the farmers. As suggested by Jähn et al. (2006), they recognise that the value chain is of great importance to their business, but they do not recognise that they may benefit from improvements by their suppliers in order to confront increasing global competition. It is interesting that, for example, Findus that already acts in a global market with fierce competition is aware of their dependency to suppliers.

Overall, a lot of information concerning the products that we have studied is collected throughout the food chain. However, in most cases a limited amount of information is transferred from one actor to the next in the chain. We can only speculate on the reason for this, but apparently the actors feel that there is no need to relay information that they consider to be of no real practical use to others. Or the explanation may be as simple as a question of power. The information overhead that exists up the value chain is used to keep the suppliers in a state of uncertainty. There is, for example, no reason for Skånemejerier to collect data on which of the farmer's cows produces the most milk, since they are only interested in the milk itself. This attitude seems perfectly reasonable.

In our empirical data three drivers of inter-organisational integration specifically stand out:

- Control – Large organisations with well established brands seek control over the complete production cycle to ensure stability in quality and increase market share.
- Legislation and policies – Most of the integration stems directly from imposed requirements from authorities or interest groups.
- Economic use of resources – A limited set of integration initiatives is implemented to enable more efficient use of existing resources.

Corporate II is used to internal and external pressure (Hanseth and Braa, 2001). In our cases external pressure clearly is the more decisive force. As can be seen, the drivers are mostly of the character of being "necessity" rather than business strategic. The internal drivers are limited in our cases. Instead our internal condition seems to be of a hampering character. In our empirical data we find a relation with certain hampering conditions that seem to hinder the progress of business enhancing initiatives:

- Bargain power – The industry is dominated by a few large key actors that can set the conditions for numerous small producers. This leads to the larger player obtaining the information they want, but smaller actors having no access to data that would enhance their business.
- Organisational agendas – No common understanding of how to improve the extended value chain.
- Regarding hampering conditions, purely technical factors do not seem to be severe impediments in our empirical material. When the organisations agree upon an integration need, they seem to solve it. The key criteria concern knowing what is beneficial for business and succeeding in convincing stakeholders of the necessity.

2.6 Conclusions

ICT has developed to such a level that it is possible to create an extended value chain that integrates all steps in the chain from original producers to end consumer. However, just because it is possible to integrate does not mean companies do so.

Our study of nine companies in the Swedish food industry shows that the organisations' own agendas act as impediments, and only those companies that are large enough to exercise pressure on its partners manage to leverage the potential benefits of II integration in the extended value chain. The condition that the extended value chain lacks a common owner makes impossible some of the well established success factors for integration projects, such as top-level management support as there, basically, is no top-level management for the extended value chain. Integration is based on mutual benefits and or exercise of bargaining position of the buyer.

We see that technological innovations may facilitate integration in the future, but still the most outstanding finding from our empirical data is that organisations not see themselves as an extended value chain competing with other value chains. We see that to some extent the individual companies are dependent on the others' success, e.g. Findus needs the pea cultivator to be successful in order to continue successful business and even the larger retail chains are dependent on the success of Findus. This finding is also supported in previous research (Jähn et al., 2006). In some relations, this dependency is well recognised, in others not.

Our findings encourage more studies within the field, not least action-oriented initiatives. As we do see that bargaining power and the question of mutual benefits plays a significant role in integration of the extended value chain, we would suggest that the subject matter could be studied from an alliance-forming perspective in order to more systematically address these drivers and inhibitors. The distributed management of extended value chains is an important factor to consider. For example do some combinations of strategic IS planning characteristics seem seem effective than others (Segars and Grover, 1999), does this imply that integration initiatives directed to the extended value chain should follow a specific path due to their bottom-up nature? Many questions doubtlessly remain unanswered regarding the role of ICT in the extended value chain.

Acknowledgement: Thanks are extended to the students taking the course Business Systems (INF 653) fall 2005 (Mustansar Ali, Emma Andersson, Johan Ekelund, Sunna Gudmundsdottir, Daniel Hägg, Malin Meiby, Muhammad Naeem, Johan Modin, Mahroz Nakhaie, Magnus Olsson, Mikael Rosvall, and Leopold Schmidt) who did the empirical data collection, and to the participating organisations.

2.7 References

Browne J, Jiangang Z, (1999) Extended and virtual enterprises - similarities and differences. International Journal of Agile Management Systems 1(1): 30–36

Browne J, Sockett PJ,Wortmann JC, (1995) Future manufacturing systems – Towards the extended enterprise. Computers in Industry 25:235–254

Davenport TH, (1998) Putting the Enterprise into the Enterprise System. Harvard Business Review 76(4):121–131

Gunasekaran A, Ngai, EWT, (2004) Information systems in supply chain integration and management. European Journal of Operational Research 159:269–295

Hanseth O, Braa K, (2001) Hunting for the treasure at the end of the rainbow. Standardizing corporate IT infrastructure. Journal of Collaborative Computing 10(3-4):261–292

Henningsson S, (2007) The relation between IS integration and M&A as a tool for corporate strategy. In Proceedings of 40th Hawaii International Conference on System Science, Waikoloa, Hawaii, US

Hitt LM, Wu DJ, Zhou X, (2002) Investment in Enterprise Resource Planning: Business Impact and Productivity Measures. Journal of Management Information Systems 19(1):71–98

Jagdev HS, Browne J, (1998) The extended enterprise – a context for manufacturing. Production Planning & Control 9(3):216–229

Jähn H. et al. (2006) Performance evaluation as an influence factor for the determination of profit shares of competence cells in non-hierarchical regional networks. Robotics and Computer-Integrated Manufacturing, 2:526–535

Konsynski BR, (1993) Strategic control in the extended enterprise. IBM Systems Journal 30(1):111–142

Lubatkin M, (1988) Value-Creating Mergers: Fact or Folklore. The Academy of Management Executive 2(4):295–302

Markus ML (2000) Paradigm shifts – e-business and business/systems integration. Communication of the Association for Information Systems 4(10).

Robey D, Ross JW, Boudreau MC, (2002) Learning to Implement Enterprise Systems: An Exploratory Study of Dialectics of Change. Journal of Management Information Systems 19(1): 17–46

Schweiger DM, Goulet PM, (2000) Integrating mergers and acquisitions: an international research review. Advances in Mergers and Acquisitions 1:61–92

Segars AH, Grover V, (1999) Profiles of strategic information systems planning. Information Systems Research 10(3):199–233

Upton DM, McAfee A, (2000) A Path-Based Approach to Information Technology in Manufacturing. International Journal of Technology Management 20(3/4):354–372

Worthen B, (2002) Nestle's ERP Odyssey. CIO Magazine

Yin R, (2003) Case Study Design: Design and Methods. 3rd ed, Sage

3

Integrative Technologies in the Workplace: Using Distributed Cognition to Frame the Challenges Associated with their Implementation

Carole Groleau
University of Montréal

An important theme in the ERP (Enterprise Resource Planning) systems literature is the integrative dimension of this type of technology. In this chapter, we investigate technological integration using distributed cognition. In the framework of Hutchins (1995), founder of this approach, human action is based on the ability of human beings to integrate the various elements of the context in which they function. Drawing on Hutchins' conceptual framework, we see how the material structure of integrative technologies changes the relationship between workers and their contexts as well as how they conduct work activities.

3.1 The Integrative Logic of ERP Systems

Designated as the most popular software system of the 20th century, ERP (Enterprise Resource Planning) systems have generated much interest among practitioners as well as researchers over the last two decades (Robey et al., 2002). One of the recurring themes in the abundant literature published on the topic in popular and academic journals is the integrative quality of ERP systems and the challenges it raises during implementation (Light and Wagner, 2006).

The integrative dimension characterising ERP systems surfaced in Davenport's early writings (1998). He writes, "These commercial software packages promise the seamless integration of all the information flowing through a company—financial and accounting information, human resource information, supply chain information, customer information" (1998, p. 131).

In similar terms, the integrative dimension is discussed in various studies:

"An ERP system can be thought of as a company-wide information system that tightly integrates all aspects of a business. It promises one database, one application, and a unified interface across the entire enterprise" (Bingi et al., 1999, p. 8).

"Indeed, this comprehensive packaged software solution seeks to integrate the complete range of a business' processes and functions in order to present a holistic view of the business from a single information and IT architecture" (Klaus et al., 2000, p. 141).

In these definitions, integration is used to explain how the information stored by the software brings together various work processes into a single logic. It is this form of integration that we investigate[1]. More specifically, we examine how these technologies can take part in the work practices of organisational members, knowing that these tools, as Davenport (1998; 2000) argues, are sometimes rigid and incompatible with existing work arrangements: "[A]n enterprise system, by its very nature, imposes its own logic" (1998, p. 122). The difficulty in bringing together different logics, for people using these technologies, has been associated with numerous cases of ERP implementation failures (Wagner and Newell, 2006).

Organisational transformation following the arrival of an integrative technology is often examined through the technological potentialities it renders accessible to organisational members. These technical characteristics become the determinant of the changing work practices that follow from its use. An important stream of research has been criticised for favouring technology over other dimensions in their analysis of organisational practices following the implementation of ERP systems (Hanseth and Braa, 1998; Rose et al., 2005; Botta-Genoulaz et al., 2005).

Apart from these studies focusing mainly on technology, other researchers, inspired by the social sciences, have explored alternative frameworks for investigating the arrival of integrative technologies in organisational settings (Robey et al., 2002; Kraemmergaard and Rose 2002; Cadili and Whitley, 2005; Elbanna 2006). Within this research, the technological dimensions of ERP systems have tended to be downplayed. As Rose et al. (2005) argue: social constructionist IS theorists display "paradigm consolidation" in underestimating the influence of technology. It has become a norm to focus on the actions of humans, and a kind of heresy to point to the effects of technology. We challenge social theorists also to be specific about what the technology does." (Italics in original, p. 147)

We want to address the challenge raised by Rose et al. (2005). Apart from providing a new framework for social constructionists, we feel our work can also offer an interesting research avenue for the first stream of researchers who have focused on the specific characteristics of integrative technology at the expense of human actions in their analysis: "many articles from the "engineering community" emphasise the drastic importance of human factor for the implementation or adoption of ERP processes, without really being able to go much further in that direction (Botta-Genoulaz et al., 2005, p. 519).

---

[1] Within the ERP literature, integration can also be framed as the capacity to adapt different technologies to one another, including various modules within ERPs or ERPs with other technologies (Davenport, 1998).

To do so, we will rely on distributed cognition (Hutchins, 1995). Like other approaches, such as situated action (Suchman, 1987) and activity theory (Engeström, 1987), distributed cognition is associated with the workplace studies movement, which aims to "address the social and interactional organisation of workplace activities and the ways in which tools and technologies, ranging from paper documents through to complex multimedia systems, feature in day to day work and collaboration" (Heath et al., 2000, p. 299).

Among the approaches within this movement, distributed cognition distinguishes itself because of its conceptualisation of artefacts, which allows the researcher to analyse in detail the specificities of tools such as integrative technologies without focusing exclusively on them (Groleau, 2002). More particularly, our study will develop the concept of artefact syntax introduced by Hutchins (1995) to address the way in which work environments and artefacts are mutually constituted.

In the following section, we will explain the tenets of distributed cognition focusing on the artefact syntax, which we will subsequently apply to a case study to investigate the worker's experience of integrative technologies and the challenges associated with them.

3.2 Distributed Cognition: A Framework to Study Integrative Technologies

Distributed cognition is associated with the convergence of three movements: the increasing popularity of authors such as Leont'ev, Vygotsky, Dewey, and Wundt, the development of situated cognition (Lave, 1988), and the challenge of conceptualising human–computer interactions (Salomon, 1993; Rogers, 1993; Rogers and Ellis, 1994). To explore this framework, we draw from the work of Hutchins (1995), who is considered the founding figure of distributed cognition.

Researchers interested in distributed cognition study the material and social conditions under which actions take place. They want to expand the notion of cognition, traditionally defined as solitary mental activity, by dissolving the boundaries of the human body to be able to conceptualise cognition as a series of interactions among media located inside and outside the individual's skin.

In this framework, human action is based on the ability of human beings to integrate the various elements of the context in which they function (Hutchins, 1995). The context is defined as a set of structures of material or social origin, from which individuals draw the information necessary to undertake action. As such, cognition is distributed to the extent that it draws on a variety of structures external to the human body.

One important work within the distributed cognition movement is Hutchins' empirical study on marine navigation. More specifically, Hutchins studied the evolution of artefacts used in marine navigation over the past centuries. He explores how the regularity of natural cycles was harnessed and conveyed by a variety of tools. More specifically, he studied Micronesian peoples who used no artefacts to get their bearings when at sea. These navigators measured the distance and the position of their small craft by using natural landmarks such as the position of surrounding islands and the movement of celestial bodies. According to Hutchins, these natural points of references can be found in the first artefacts created for navigation. Thus, he explains how in an artefact such as the astrolabe[2] a series of icons is combined, which represent regularities of the natural world such as the movement of the sun and stars in order to aid humans to navigate the seas. These icons, which illustrate the different cycles of the natural elements, are translated, over time, into alphanumeric codes. Thus, today, navigation instruments are encoded with numbers, among other things, for situating a craft according to its latitude and longitude.

In this study, Hutchins (1995) analyses the shift from natural to symbolic regularities. In doing so, he introduces the syntactical dimension of artefacts:

> The regularities of relations among entities in this world are built into the structure of the artefact, but this time the regularities are the syntax of the symbolic world of numbers rather than the physics of literal world of earth and stars. The representation of symbolic worlds in physical artefacts, and especially the representation of the syntax of such a world in the physical constraints of the artefacts itself, is an enormously powerful principle (Hutchins 1995:107).

Beyond providing the means to analyse the historical development of artefacts for a given set of activities, we feel the artefact syntax offers an interesting conceptual lens to study the implementation of integrative technologies in organisational contexts.

Fundamentally, Hutchins' concept allows reflection on how regularities are integrated into artefacts that support collective action. We believe that in the life of current organisations, these regularities aren't necessarily drawn from the natural world. Recognising and identifying what constitutes regularities linked to an activity so as to render them material is in itself a complex question. The work at hand, as well as the attributes and cognitive needs related to this work, are not, in our opinion, objective realities that are self-evident, but rather a reality that begs to be intersubjectively defined in a context where the interests of each influence their perspectives on the work to be carried out[3].

In studying artefact syntax within another type of work environment, we must question ourselves on the nature of the regularities. Furthermore, we need to explore the syntactical dimension of material entities by investigating the manner in which the artefact integrates, conveys, and juxtaposes the regularities of the environment in which it is installed. Hence, by manipulating the artefact conjointly with other already present artefacts, it is possible to examine the enablements and

---

[2] "The astrolabe a portable mechanical model of the movements of the heavens, was invented in Greece around 200 BC.... It is a sedimentation of cosmic regularities. The astrolabe also enables its user to predict the positions and movements of the sun and stars" (Huchins, 1995:96-97).
[3] The issue of the political dimensions of technological design was the subject of a stimulating debate between Suchman (1994) and Winograd (1994).

constraints inciting different forms of action that are rendered possible by this environment.

We analyse how workers reconfigure their work practices with the implementation of an integrative technology considering the type of regularities it contains and the way in which they are configured within this particular category of artefact. We would like to address the following questions: What is the source of the regularities that circulate within the technological artefact? How do the regularities and their configurations within integrative technologies differ from the artefacts previously used in the work context? How do workers reorganise their work practices using this new technological artefact?

3.3 Mutating Artefacts, Mutating Work: The Case of Billing Services in a Hospital Environment

3.3.1 Some Methodological Points of Reference and a Description of the Visited Site

Our empirical investigation focuses on the computerisation experience of accounting clerks within a hospital facing the implementation of a new technology centralising patient files. This willingness to share common data and practices within one technological solution was the motivation behind the implementation of this new software. More specifically, the chosen technology, COMPTA, was to bring together two already existing information systems, a medical-administrative system, MEDIC[4], and an accounting system, FINATECH. The MEDIC system allows the management of all information concerning a patient's registration at the hospital as well as all of his or her movements once inside the hospital (e.g., change of sector, room type, etc.). The FINATECH system organises the financial activity register of the hospital.

We feel the studied technology meets the criteria of an integrative technology because it merges two existing logics, the administrative and accounting work processes, in one central technological system. It merges different aspects of the patient profile in a series of electronic documents used by the accounting clerks as well as other workers such as the admission personnel. Our study will adopt the point of view of the accounting clerks.

The accounting clerks worked within billing services in a university hospital in Quebec, with a bed capacity of 510 beds. This hospital employs more than 3,600 health professionals grouped in seven large departments: (a) teaching, (b) professional and hospital services, (c) nursing care, (d) planning and communication, (e) human resources, (f) financial research and resources, and (g) technical services. Billing services is located in the financial resources and the technical services departments.

---

[4] The names of the information systems have been changed in order to protect the confidentiality of the billing service.

Its mission is to maintain billing, accounting, follow-up and collection of accounts receivable, in collaboration with all sectors of the hospital. Billing services is divided into four activity sectors: outpatients, regular inpatients, housing or accommodations, and service sales. Our research focuses specifically on the accommodations sector that manages the accounts of long term care patients. This sector employs two clerks, named Dominique and Evelyne[5], who manage accounts receivable.

To glean the data necessary for our case study, we called upon several data collection methods, including document analysis, interviews, and observation. As a first step, document analysis allowed us to sketch of an overall picture of the software installation process. We consulted with many reports and documents related to the computerisation[6] project and billing services, including a complete dossier on the computerisation project (which included the document presented to the head of finances, the results of a time and movement study carried out by the administration technician of the billing services employees, the minutes from all the computerisation committee meetings, and the final computerisation report), the hospital's annual report, and a variety of previously used forms.

The documentary research was simultaneously carried out with the first semi-directed interviews. During these interviews, we met with the two heads of the department, the head of services and the administrative technician. We discussed with them the role of the billing services department, its performance, its links with the other departments, its computer system, and the computerisation project.

Finally, a large part of our data came from observations of the billing services employees, Dominique and Evelyn. Over the course of more than two months and on a daily basis, we were able to explore the everyday work of these two clerks in action in their environment and, thus, to gain access to data that are difficult to obtain through interviews.

Our observations were largely concentrated on these two clerks in billing for long term care accommodations. Their principal tasks were to collect the necessary information for calculating the accommodation rate for each patient and to follow-up on case files and payments. We will describe in detail in the following sections the nature of their work and the artefacts that they use to accomplish it.

3.3.2 The Computerisation of Account Billing: The Evolution of Documents and of Clerical Work

3.3.2.1 Evolution of the Documents Used for Managing Patient Accounts
At the time of our observations, 115 patients were staying in the long term care unit at the hospital studied. The patients in this sector are rarely admitted directly to the long term care unit. They are first hospitalised for a health problem and are subsequently transferred to a long term care accommodation unit. The clerical

---

[5] The names of the participants in our study have been changed to protect their confidentiality
[6] In our text, the computerisation project or computerisation refers to the installation of the COMPTA software.

work consists of finding out the patient's choice of room type and in managing the patient accounts. To do this work, a series of different documents was produced before and after computerisation.

Before computerisation, the most important document for taking account of the financial activities associated with a patient's stay in the hospital was a yellow-coloured accounting card, commonly called the yellow card[7] by hospital employees. When a person arrived at the hospital, the admissions staff created a file in the MEDIC software and issued a yellow card in the new patient's name. The yellow card was then sent to the billing officer assigned to the type of accommodation required by the patient. The yellow card followed the patient through all of his or her movements in the hospital. So, when he or she moved from inpatient care to longer term care, the patient's yellow card was transferred from the inpatient billing clerk to the long term care billing clerk.

The billing employees based the financial profiles they created for patients on these yellow cards, which were paper documents on which a range of information was recorded. On one side, the card was organised with a series of headings where the hospital personnel looked to note the identity of the patient, the locale of where the patient was staying in the hospital, the rate for the room that was being occupied, the credit and debit information for the billing and payment for the room occupied by the patient. On the other side of the card, a list of room types was printed. In Quebec hospitals, three types of room are available: public, semi-private and private. The patient or his or her advocate checked the choice and signed the card. A series of lines was also printed on the card, and this space was used by members of the clerical staff to record the follow-up of different steps. These commentaries were either recorded by hand or typed, depending on the clerk's choice. All other documents involved in case file management (forms or billing receipts, statements of account) were stapled to the yellow card. These cards were stored and sorted by alphabetic order in paper files.

Before the arrival of the COMPTA software, which was intended to link FINATECH and MEDIC, billing services used two terminals to access the case files of patients on MEDIC. It was under-consulted, however, because all the case file information on MEDIC could be found on a patient's yellow card or the documents attached to it.

Since computerisation, the yellow card is no longer in use. The information once included on this card was reconfigured into a series of electronic documents presented visually, one by one, on a computer screen. The data is thus visually organised on different screens: the identification of the patient, his or her movement in different hospital rooms, the dates and amounts of payment, the amount owed to date by the patient, and the transfer of money from one account to another when the patient moves, for example, from active hospitalisation to long term care.

All employees of the billing services department consult the different computer documents produced by the COMPTA software from the computer installed on their desks. A paper file completes the documents integrated into the computer

---

[7] Hence, this is the term we will employ in the rest of this text.

program, and these electronic documents are never printed. The paper file contains statements of account, billing receipts, room choice form, and all other documents worth saving are stored.

Two paper documents have remained unchanged since computerisation. One of them is a hospital form for the patient's choice of room type. On one side, the list and rates for different rooms are outlined. The patient must check one of the boxes next to the different choices offered. A space is also reserved for the patient's signature on this form. On the other side of the form, another list of room choices and rates is provided for patients who are not residents of Quebec. Once again, the form contains spaces for indicating one's choice and for signing. The second document is a governmental form, filled in by clerks and subsequently forwarded to the government's housing financial aid service to determine accommodation costs for Quebec residents based on their personal information and financial resources. Both forms were stored in paper files before and after computerisation.

3.3.2.2 Opening and Managing the Accounts of Patients in Long-term Accommodation

The work of the clerks Dominique and Evelyne begins when patients are admitted to long term care.

For each new patient housed in long term care, Dominique and Evelyne perform what they call an inquiry, which consists of communicating with the patient's family or with the patient him- or herself in order to obtain information about the patient's insurance company (if any exists), the mode of payment for hospital fees, and the type of room desired. This information will be used to complete the form for the choice of room type and to calculate the room rate. The accommodation costs are established by the government's housing financial aid service following an evaluation of the patient's financial situation and of the type of room occupied. Remember that the rooms offered are either public, semi-private, or private. The cost of accommodation, according to the room type chosen, may be entirely covered by the Quebec government. This is the case for residents of Quebec who opt for a public room. If, on the other hand, a patient chooses a private or semi-private room, he or she must pay an extra charge. This amount may be paid by an insurance company, if the patient has a policy that covers this type of claim. If not, the patient must cover the expense him- or herself. The long term care billing clerks must thus find out if the type of room chosen by the patient, or by his or her family members if he or she is incapable of making this choice, and must obtain the relevant billing information, be it an insurance company or an individual.

Inquiries can drag out over several days, if not several weeks, depending on the complexity of the situation and the rapidity of returned calls from the contacted family members. Previously, to help find their bearings during the effort to obtain the room type selection and the coordinates of those who will assume the financial burden associated with this choice, the two clerks previously wrote a series of notes on the back of the yellow card. Here, they generally noted the date and the name of those with whom they spoke with or all other contextual information that would help them follow-up on the inquiry. Since computerisation, the clerks attach an adhesive blank sheet of blue paper to the paper case files where they write down the contextual information of the inquiry where they generally used to record the patient's name, date of admission, the wing of the hospital where the patient is staying, his or her room number, and the type of room choice. They have developed their own abbreviations and codes, for example, highlighting with a coloured marker the elements that seem important. When the clerks have all the necessary information in hand, they complete a form and send it to the government so that the accommodation costs may be calculated. This form has stayed the same before and after computerisation.

As we mentioned earlier, the patients staying in long term care have usually been hospitalised in another care unit in the hospital before transferring into long term care. Before the arrival of COMPTA, the patient kept his or her yellow card, regardless of his or her movements in the hospital. For example, in the case of a patient who had already occupied a room in the hospital and who was transferred to the long term care wing, Dominique and Evelyne collected the yellow card used by the inpatient employees. They then followed the trail in the accounting section of the card to determine the start of the patient's hospital stay.

Since the integration of the COMPTA software, Dominique and Evelyne must also open a new paper case file each time a patient is admitted to long term care, whether the patient comes from another unit in the hospital or not. Moreover, in the software, a new case number is assigned and a series of electronic documents are created for this new case. If the patient has been transferred from another hospital unit, certain personal patient information is transferred from the old case, such as name, date of birth, name of spouse, address, telephone number, health insurance number, and emergency contact person. The financial data relative to the hospital stay before the patient's arrival at long term care are not transferred, however, in the ensemble of the computer documents grouped under the new case number. Dominique or Evelyne opens a new account and calculates the accommodation costs starting from the date of the transfer.

3.3.2.3 Collecting Accommodation Fees

The accounting services department follows a calendar of 13 annual financial periods. However, the government requires monthly calculation (i.e. 12 annual financial periods) of the accommodation fees for patients of long term care units. With the arrival of the new software that follows the annual system of thirteen 28-day periods, a conflict arises between the two modes of time calculation.

An important part of the clerical work consists of ensuring that the accommodation fees are billed to and paid by the patient, by his or her family, or by the insurance companies. This means that the monthly preparation and sending of account statements must be done on the first day of each month. Because it is automatically deposited with the hospital, the part of the accommodation cost covered by the government is not sub
ject to monthly invoicing.

Before computerisation, account statements were prepared by the clerks using the appropriate forms on which they would type the amount due, next to which they wrote the months and days covered by a given account statement. This task was generally spread out over three days. To produce the account statement, they consulted the yellow cards to see if the previous month's balance had been paid and to verify if a full month's accommodation ought to be charged. They then typed the statement of account, noting the corresponding accounting comments on the yellow card. The amounts normally charged to patients for the three different room types were an important point of reference for the clerks. This amount effectively served as a code permitting them to rapidly decipher each patient's situation.

Now, the production of account statements is automated, but nevertheless requires particular handling to respond to the specificities of the civil calendar. First, the clerks consult onscreen the electronic document showing the amounts due. These are normally displayed according to financial period. Thus, it is common that a given month will overlap two financial periods. To ensure that the amounts on the screen actually correspond to a month of accommodation, the clerks use a calendar and a calculator to multiply the number of days the patient stayed in the long term care unit by the daily rate for the chosen room type. This calculation serves to verify the reliability of the numbers displayed on screen. Each situation is unique with the patients in long term care, there can often be patient moves to other care units or departures. If the amount on the screen corresponds to the one calculated using the daily rate, then the clerks can issue the account statement. When the amounts do not correspond, they retry using different mathematical manipulations of the days and the rates to understand where the sum shown on screen came from. This can also require that they consult a table of the patient's moves in the hospital, other computer documents, or the contents of the paper case file. Once the clerks understand the amount shown on screen, they are ready to issue the statement of account.

The creation of a monthly account statement, however, requires a rather particular manoeuvre. Because the 28-day periods never correspond with the months, the clerks take the necessary steps with the computer program to indicate the patient's departure from the hospital. Although the patient doesn't really leave the hospital on the first of every month, by indicating this departure, the clerks are able to print an account statement on paper, taking into consideration the amounts owed on the first of the month. After the account statement is produced, the case file for the patient (who never really left the hospital) is immediately reactivated. The account statement produced at the end of this series of operations indicates the name of the addressee and the amount due. In order to avoid billing confusion on the part of the patient, his or her family, or his or her insurance company, the clerks type on the computer printed document the details relative to the months or days covered by the account statement that will be sent out.

Finally, the new software's arrival signalled the elimination of one last step in the clerical work. Before computerisation, the clerks also had to fill out paper forms associated with accounting ledgers by writing the specifics of each of the financial transactions so that these would be entered in FINATECH to produce financial statements. This step is no longer necessary as the COMPTA and FINATECH programs are linked to one another.

3.4 From Paper to Screen: Analysing the Change that Organisational Members Experienced

Now that we have described the operation of the clerical work as well as the amplitude of the changes brought about by computerisation, we will now analyse this data in order to answer our research questions[8]. We seek to understand the nature of the regularities as well as the manner in which they are configured in the documents conveyed by paper and computer documents. We will employ the concept of artefact syntax to grasp the dimensions of change associated with the arrival of the electronic documents produced by the COMPTA program.

3.4.1 Articulating Regularities Within the Artefacts Themselves

Starting from our empirical data, we have attempted, like Hutchins did for his study on navigation, to characterise the nature of the work as well as the regularity of the artefacts used in the carrying out of everyday activities.

The clerical work consists of taking note of the type of accommodation chosen by the patient as well as the parties who will pay for this service, sending out account statements, receiving funds to pay for accommodation, and finally, keeping records of all of these operations. To identify regularities, which in the case of the clerks are not tied to repetitions of natural cycles, we drew on documents from before and after computerisation in order to identify the criteria chosen for bringing together this information.

Thus, the ensemble of the clerks' activities and the regularities follow a transactional logic where two parties enter into relation with each other to carry out an exchange of goods or services for financial compensation. The regularities brought together in these artefacts answer to a series of questions tied to the characteristics of a transaction: Who? What? How much? When? More specifically, these regularities allow us to take into account the identities of the parties engaged in the transaction as well as the nature of what is exchanged and the frequency of these exchanges.

In this particular case, the *who* refers to the different parties that take part in the transaction. The financial compensation for the service of accommodation that may be paid by one or several of these parties: the government, the insurance company, the patient or a family member who may act on his or her behalf. The clerks working for long term care accommodation do not handle transactions with the government, as previously explained, but they must still manage the transactions with the insurance companies, the patients, and/or their family members.

The *what*, or the object of the transaction, is the type of accommodation for each patient. Here again, several alternatives are possible: a public room, a semi-private room, or a private room. The *how much* corresponds to the financial remuneration that will be offered in exchange for the accommodation service.

---

[8] A preliminary data analysis of this case study was published in Communication and Organisation Vol. 33.

Finally, the *when* also constitutes a dimension that reveals more than one criterion for measuring time. The rates and payment due dates are organised monthly like rent, while the monitoring of funds paid by the different parties is recorded in hospital documents according to an organisation of annual time into 13 periods of 28 days each.

3.4.1.1 The Transaction Broken Down: Who Paid What? When?

When we studied the regularities in question from the different artefacts, we first remarked that the same regularities circulated before and after computerisation. However, if the regularities have remained the same, the manner in which they merge with the artefact differs since the arrival of the integrative technology. For example, we gathered many comments from our clerks confronted with the substitution of the paper artefact by the electronic one:

"When you looked at it (the yellow card), you would know right away"

"It wasn't complicated; everything was on the yellow card"

"The yellow card got us through everything"

"The yellow card is more concrete, it's not abstract"

"(Since computerisation) You always have to go back and forth from one screen to another (tableau)…yes, and sometimes, you've just had it up to here!!"

In sum, while on the yellow card, the dimensions of the transactions are juxtaposed one next to the other, in the computer documents, the regularities characterising each patient's account are dispersed over several electronic documents.

The yellow cards offered simultaneous access to all information about the financial transactions between a patient and the hospital so that with one glance a clerk could grasp all of the transaction's points of reference, thus offering an action horizon that differs from the electronic documents, which don't allow a simultaneous overview of information such as the identity of the parties engaged in the transaction, the room type, the due dates, and the amounts and dates of payments on the patient's account. More specifically, the *who, what, how much,* and *when* that were previously presented side by side on the same paper document are now dispersed over different electronic documents that are impossible to examine simultaneously.

This new configuration of regularities within the electronic artefact causes problems for the daily clerical work activities. The clerks encounter these difficulties, for example, when they must produce a final account statement for a patient who has just passed away and who had insurance. One glance at the yellow card would have sufficed to capture all the information necessary: the date and amount of last payment, which allowed the clerk to determine the number of days and the amount to invoice. The yellow card also included the coordinates of the insurance company. To obtain this same information with the COMPTA program, the clerks must consult at least three distinct electronic documents.

This part of the analysis was devoted to identifying regularities, but more importantly, to the manner in which they interrelate within the studied artefacts. We took up again the concept of syntax, introduced by Hutchins, to examine the different relationships of interdependence between the regularities conveyed by the paper and electronic artefacts. Our data illustrates that the artefacts, before and after computerisation, juxtapose and combine regularities differently, offering a different relationship with the environment and a different action horizon since the technological change took place.

Yellow card   Computerised
                                                    documents

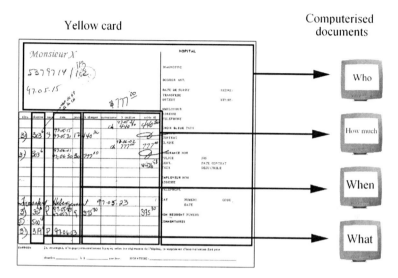

Figure 3.1. The organisation of regularities within the yellow card and the computerised documents

3.4.1.2 28, 30, or 31 Days? The Temporal Organisation of Financial Transaction Management

While in the preceding analysis, we saw the rise of new ways of organising the diverse regularities of financial transactions in the composition of the two types of documents, the discussion that follows takes a more specific approach to the different temporal regularities of the paper and electronic artefacts.

Traditionally, the two calculations of time (by 28-day period and by month) have coexisted in the work world of the long term care clerks. Although each of the two methods for compiling the days of patient accommodation has different regularities (the month and the period), there was never really any conflict between the two before the arrival of COMPTA. The monthly calculation required by the government for calculating accommodation was used on the yellow card, which was, as we may recall, a key paper document for the long term accommodation clerks before computerisation. The calculation by period was systematically used for all accounting documents, including the accounting ledgers used by the accommodation clerks to record different financial transactions. The clerks had only to write in the amounts invoiced or paid with the date and the coordinates of the account in the accounting ledgers. Hence it is at the moment of electronic processing of these accounting ledgers that the data was compiled for producing official documents following the criteria of periods with FINATECH.

With the arrival of the COMPTA program, the two temporal regularities are in constant confrontation with each other in the framework of the clerical work. By integrating one of these temporal units into the program, the information conveyed by the electronic documents is expressed exclusively according to this unit, which is the 28-day period. The organisation of the year into 13 periods of 28 days each responds to the requirement for producing accounting documents, such as financial statements. However, this breaking down of time by period is not what the clerks mobilise when billing and collecting accommodation fees.

The choice to privilege one of the two temporal regularities in the production of electronic documents has had the repercussion of orienting all readings of patient accounts through this temporal filter. Concretely, the materialisation of this temporal scale in the electronic artefact obliges the clerks to constantly perform an exercise of translation by using a calendar and a calculator to transform the financial data organised by period into a monthly logic.

However, as we noted in the case description, the most striking strategy remains the indication in the COMPTA program on the first of each month that the patient has departed the hospital in order to produce the account statements. This operation, however strange we might find it, allows for a new temporal organisation for the calculation of the patients' hospital stays. The indication of departure is, in effect, a very drastic method for marking the end of the month and for thus re-establishing the monthly calculation necessary for producing account statements.

3.4.1.3 From Yellow Card to Blue Paper: Patient, Where Are You?

Until now, we have focused our analysis on the juxtaposition of different regularities as well as on the multiplicity of organisational criteria that may coexist within even one of these regularities. We have seen in the two examples that the arrival of a new artefact altered the clerk's relationship with her work. In this section, we will continue to explore the hierarchical organisation of transaction data in the newly implemented integrated technology.

Before computerisation, the yellow card influenced the accounting process by defining the patient as a unit of work. The patient always kept the same card, no matter the wing or care unit he or she stayed in. In billing services, the clerks handed over the yellow cards of patients who transferred from one care unit to another. The same document was thus used to manage the account of the patient who was transferred, for example, from inpatient to long term care.

After computerisation, the documents produced by the software are organised by case. A case is defined by a patient's stay in a particular care unit. When the patient changes units, for example when he or she moves from inpatient to long term care, a new series of electronic documents bearing a new case number is produced. Thus, the patient's hospital stay is broken up into multiple cases if the patient changes care units during his or her hospitalisation. Concretely, this means that, with the exception of the table showing patient movement in the hospital, the information made accessible on each electronic document only partially explains the patient's stay in the hospital. In order to harmonise with the technology's logic, a new paper file is created each time a new case is created. The historic dimension of the patient's stay, relating his or her path through the different hospital units, is thus lost in the case-by-case organisation of the electronic support documents.

It is interesting to note that case numbers existed before computerisation and that these were also replaced when a patient changed units. The case number was recorded on the yellow card and when this number changed, it was written next to the previous number, even if the account management took place on the same yellow card.

While accounts may be hierarchically organised according to different criteria, one can see here a change from the order of *who* to the order of *what*, or more precisely, from the patient engaged in the financial transaction to the type of service offered, such as accommodation for this patient in different care units of the hospital. In our opinion, this new hierarchy of the data that is associated with integrative technologies has strongly contributed to the feeling of the "loss of the patient" expressed by the clerks during our study: "The yellow card made them (the patients) more human." Indeed, the yellow cards were more than simple vehicles carrying a series of data about transactions; the clerks projected onto these artefacts the actual patients and their histories. For them, to enter into a relationship with the artefact was to enter into a relationship with the patient. This connection with the artefacts as vehicles representing the patient disappeared with the arrival of the new technology.

The two accounting clerks were not immobilised by the deficiencies uncovered during this study. Indeed, following COMPTA's installation, they invented an artefact to overcome the constraints of their newly modified work environment: a blue piece of paper that was completely blank. The accounting clerks attach this artefact to the paper case file documents as soon as they are created. On it, they write information about the patient and about his or her particular context. However, despite its similarities to the other artefacts in the environment, the blue paper is distinct. It is similar to the space on the back of the yellow card previously used for hand-written notes, but the blue card is different, however, from the yellow card because it does not retrace the complete history of financial transactions between the two parties (patient and hospital).

The blue paper also shares characteristics with the electronic document created for each case by the new COMPTA software in that it also does not have pre-determined fields. The clerks can write unstructured text on the blue paper, but they know that many people in the hospital have access to and consult the electronic documents, and so the clerks only write very official information in electronic artefacts in case of potential dispute or legal proceedings. The electronic document's mode of widespread diffusion and the official nature attributed to it by the clerks discourages, however, daily note-taking with this artefact. Thus, Dominique describes the data recorded on the blue paper in these terms: "Whatever is of no consequence...no...whatever is not important...no...in a word, you know...whatever is day-to-day."

We believe that this new artefact carries out the important function of reintroducing the patient into these clerks' work process. The patient who previously appeared on the yellow card now takes the form of the blue paper and still occupies an important place in the clerks' informational universe, even if the blue paper contains data about the patient that also ends up in different electronic documents. Aware of the redundancies created by this new artefact, the clerks nevertheless appreciate that it offers easy access to the patient's profile: "The blue paper is just the icing on the cake. If I lose it, it's not a big deal; I transcribed it all on the SR-80 (governmental form) and on the screen."

Using the concept of tool syntax to analyse our data suggests different articulation formats for the regularities within the artefacts studied. The way in which these regularities materialise in different combinations alters the relationship between the clerks and their work environment. From a world where the whole of a transaction was easy to read and carry out, and where the yellow cards gave form to the patient, the clerks now find themselves in a world of electronic documents that represent in a discontinuous fashion both the different dimensions of the transaction as well as the patient's stay in hospital. In this context, it was essential for them to undertake a series of actions, such as finding other points of reference or creating a new artefact, in order for them to be able to make sense of their tasks as a function of the particular work environment available to them since the change in document support.

3.5 The Contribution of Distributed Cognition to the Study of Integrative Technologies in the Workplace

As we will argue, our use of distributed cognition, focusing particularly on artefact syntax concept, allows us to develop a new frame for understanding integrative technologies as they are implemented in the workplace. We feel our study helps to clarify this phenomenon by making a series of observations: (a) Regularities contained in artefacts used in the workplace are drawn from recurring practices of a variety of collective entities such as professional groups, organisations, and society; (b) within integrative technologies, regularities are hierarchically ordered, standardising the outlook on work processes; and (c) artefacts and situations are mutually defined through human decisions. We will now develop each one of these propositions, following our fieldwork.

3.5.1 The Nature of Regularities Circulating in Artefacts Used in the Workplace

Starting from observations of the work of the clerks, we were able to associate these regularities with a number of sources. As the organisation that we studied is a hospital regulated by the government, many regularities stem from the norms of the ministry of health that oversees hospital administration. More specifically, this is the case for the price of rooms, the types of rooms, and the choice of parties participating in covering the cost of the room.

On the periphery of organisational logic, we also note that professional practices also constitute another source of regularities in the work environment studied. For example, organising time by 13 periods of 28 days each is an accounting norm that allows one to break up the 365 days of the year into equal periods that are more easily compared. The professional practices in accounting also showed us another type of regularity that goes beyond the simple choice of *who, what, how much,* and *when.* For example, the accounting ledgers filled in by the clerks before computerisation organised information in a series of different columns in a pre-determined sequence according to the rules of accounting.

Finally, Western culture brings with it another organisation of time that is traditionally used for the payment of rent.

To these regularities, with origins in organisational, professional, and cultural logics, we add other regularities generated by the observed workers to help orient themselves in carrying out their activities. For example, when creating a blue paper, the clerks chose repeatedly the same type of data to describe the patient. Moreover, as the blue paper is blank, the regularities manifest at both as content and also as data that is sometimes underlined, other times highlighted. We note that these regularities emerge from the work practices of the clerks rather than from imposed norms, whether organisational, professional, or cultural.

The origins of regularities was briefly addressed by Hutchins (1995) who recognised that culture influences the constitution of artefacts:

A way of thinking comes with these techniques and tools. The advances that were made in navigation were always part of a surrounding culture. They appeared in other fields as well, so they came to permeate our culture. This is what makes it so difficult to see the nature of our way of doing things and to see how it is that others do what they do (Hutchins, 1995:115).

We build on his work by identifying more specifically how these regularities are configured in artefacts such as integrative technologies.

3.5.2 The Syntactic Organisation of Regularities Within Integrative Technologies

The focus of our study was the implementation of integrative technology. The value of using the concept of artefact syntax was the potential it offered to show how these regularities were combined in material form in the work environment we studied.

By comparing the yellow card with the new electronic documents, we observed that the different dimensions of the transactions moved from being juxtaposed in one paper document to being scattered across a variety of electronic documents. This change in the configuration of data is not necessarily associated with integrative technology but still represents an important barrier for users in the conduct of their daily activities with the new artefact (Groleau and Taylor, 1996).

Beyond juxtaposing the elements of transaction into a new pattern, the new software ordered data previously contained on the yellow card in a hierarchical fashion. First of all, time—which had previously been expressed in the logic of calendar months as well as in the logic of accounting periods on the yellow card— was now exclusively organised along accounting periods. The two criteria previously used were useful because they allowed clerks to draw from the yellow card the necessary information to write in accounting ledgers as well as to prepare monthly statements. As we saw in our case study, to overcome the imposition of a criterion that does not fit with their activities, workers either translated data from one logic to the other, in order to make sense of it, or they developed a stratagem to impose their own logic on the artefact.

Second, the criteria structuring the whole set of data stored in hospital files shifted from a patient logic to an accommodation logic. This new data configuration made it difficult for clerks to grasp the patient profile, especially at the time of our inquiry. Again, the clerks acted on this problem by inventing a new artefact, the blue sheet, to reintroduce the patient in their work process.

From our analysis, we see different levels of data organisation within the artefacts framing the way clerks approach their work practices. Unlike juxtaposition, the hierarchical organisation of data along certain criteria at different levels imposes on its users a common frame for understanding data. In doing so, the technology meets its objective of standardising along one process the work practices of those working with the technology. Concretely, it means that the choice of data organisation standardises users' outlook on the work process. As illustrated in our case study, the implementation of integrative technologies raises questions regarding the compatibility between the criteria chosen to hierarchically organise data contained in integrated technologies and its compatibility with the whole set of activities performed locally.

In the previous section, we saw that regularities are drawn from various norms associated with collectives such as professional, organisational, or even social entities. These regularities coexist and confront one another in the choice of criteria used to standardise practices through material artefacts such as the integrative technology we have been studying. In our example, difficulties arose from the use of regularities associated with the practice of accounting to organise the set of tasks performed by the clerks.

Our discussion has focused mostly, up until now, on technological characteristics of integrative technologies. But, in each of the examples presented in this section, we concluded by explaining how organisational members overcame the technological difficulties by coming up with a variety of inventive solutions. We will continue our discussion by emphasising the interplay between technological characteristics and human intervention to see how they come together to shape emerging work practices in newly computerised environments.

3.5.3 Artefacts and Situations Are Mutually Defined Through Human Decisions

Although we have insisted in our analysis on "what technology *does*", we see situations and artefacts as mutually influencing each other. We can see from our discussion of the case study how artefacts contribute to shape workers' relationship to their work environment. But, this relationship is not unidirectional. Our data also illustrates how situations can lead to the emergence of new artefacts. It was the case of the blue paper, created by workers, to circulate data that had become invisible to them since computerisation. We can argue that situations and artefacts mutually constitute themselves, as they both evolve at their own pace.

Artefacts and situations evolve through human decisions as organisational members attempt to circumscribe them. The creation of new artefacts, whether it is emergent like the blue paper or planned like the software package that was implemented in the hospital after a long decision-making process, raises a series of questions regarding which data it will render accessible to its users as well as the way this data will be juxtaposed, organised and hierarchised. As argued by Suchman (1994), these decisions become political as choices of grouping and categorisation are undoubtedly linked to the exercise of power. We agree with

Suchman that technological projects are often politically charged and a means to exercise discipline. Still, we think workers, as well as managers, involved in computerisation are often not aware of the cognitive challenges associated with the arrival of a new artefact.

In the visited hospital, a careful study of both clerks' activities was performed by the accounting managing team in order to better plan computerisation. Their study clearly identified the time spent doing each activity without really considering the environmental resources, such as other artefacts, used to perform them. This example might not be representative of all computerisation practices but they focus on work as a series of planned actions without considering the context in which these actions unfold. This method relying on planned actions has been largely criticised over recent decades (Sucham, 1987; Sachs, 1995).

We believe the use of distributed cognition allowed us to conceptualise artefacts as material entities enabling and constraining human activities through their specific characteristics. We feel their durable and material form limits their interpretive flexibility, which we think is not exclusively determined by the humans manipulating them, as some researchers have proposed (Boudreau and Robey, 2005). Still, we argue that artefacts are material entities produced by humans, carrying with them a view of the world which we can glimpse through the regularities that are ordered, more or less intentionally, within them.

Acknowledgements: I would like to express my gratitude to Nicole Giroux who has encouraged me to write this text. I also want to thank Isabelle Guibert for collecting the data constituting the case study presented in this text and Olivia Laborde for having done some of the documentary research. I am also grateful for the help provided by Stephanie Fox in editing and preparing this manuscript in English. Finally I have benefited from the financial support of the Social Science and Humanities Research Council of Canada for its financial support (CRSH 410-2004-1091).

3.6 References

Bingi P, Sharma M, Godla J, (1999) Critical Issues Affecting an ERP Implementation. Information Systems Management Summer:7–14

Botta-Genoulaz V, Millet P-A., Grabot B, (2005) A Survey on the Recent Research Literature on ERP Systems. Computers in Industry 56(6):510–522

Boudreau MC, Robey D, (2005) Enacting Integrated Information Technology: A Human Agency Perspective. Organization Science 16(1):3–18

Cadili S, Whitley EA, (2005) On the Interpretive Flexibility of Hosted ERP Systems. Strategic Information Systems 14:167–195

Davenport TH (1998) Putting the Enterprise into the Enterprise System. Harvard Business Review July-August:121–131

Davenport TH, (2000) The Future of Enterprise System-Enabled Organizations. Information Systems Frontiers 2(2):163–180

Elbanna AR, (2006) The Validity of the Improvisation Argument in the Implementation of Rigid Technology: The Case of ERP Systems. Journal of Information Technology 26:165–175

Engeström Y, (1987) Learning by Expanding: An Activity-Theoretical Approach to Developmental Research. Helsinki: Orienta-Konsultit Oy

Groleau C, (2002) Structuration, Situated Action and Distributed Cognition: Rethinking the Computerization of Organizations. Système d'Information et Management 2(7):13–36

Groleau C, Taylor JR, (1996) Toward a Subject-Oriented Worldview of Information. Canadian Journal of Communication 21(2):243–265

Hanseth O, Braa K. (1998) Technology as Traitor: Emergent SAP infrastructure in a global organization. Paper presented at the International Conference on Information Systems, Helsinki, Finland

Heath C, Knoblauch H, Luff P, (2000) Technology and Social Interaction: The Emergence of 'Workplace Studies'. British Journal of Sociology 51(2):299–320

Hutchins E, (1995) Cognition in the Wild. Cambridge, MA: MIT Press

Klaus H, Rosemann M, Gable GG, (2000) What is ERP? Information Systems Frontier 2(2):141–162

Kraemmergaard P, Rose J, (2002), Managerial Competences of ERP Journeys. Information Systems Frontier 4(2):199–211

Lave J, (1988) Cognition in Practice. Cambridge, U.K.: Cambridge University Press

Light B, Wagner E, (2006) Integration in the ERP Environments: Rhetoric, Realities and Organizational Possibilities. New Technology, Work and Employment 21(3):215–228

Robey D, Ross JW, Boudreau, MC, (2002) Learning to Implement Enterprise Systems: An Exploraty Study of the Dialectics of Change. Journal of Management Information Systems 19(1):17–46

Rogers Y. (1993) Coordinating Computer-Mediated Work. CSCW, 1:295–315

Rogers Y, Ellis J, (1994) Distributed Cognition: an Alternative Framework for Analysing and Explaining Collaborative Working. Journal of Information Technology 9:119–124

Rose J, Jones M, Truex D, (2005) Socio-Theoretic Accounts of IS: The Problem of Agency. Scandinavian Journal of Information Systems 17(1):133–152

Sachs P, (1995) Transforming Work: Collaboration, Learning and Design. Communication of the ACM 38(9):119–124

Salomon G, (1993) Distributed Cognitions: Psychological and Educational Considerations. Cambridge, MA: Cambridge University Press

Suchman L, (1987) Plans and Situated Action: The Problem of Human-machine Communication. Cambridge MA: Cambridge University Press

Suchman L, (1994) Do Categories Have Politics? The language action perspective reconsidered. Computer Supported Cooperative Work 2:177–190

Wagner E, Newell S, (2006) Reparing ERP: Producing Social Order to Create a Working Information System. Journal of Applied Behavioral Science 42(1):40–57

Winograd T, (1994) Categories, Disciplines, and Social Coordination. Computer Supported Cooperative Work 2:191–197

4

ERP Implementation: the Question of Global Control Versus Local Efficiency

Anne Mayere, Isabelle Bazet
University of Toulouse 3, LERASS

4.1 Introduction

This chapter addresses the problem of implementation of an ERP system, and more precisely the adaptation of the system to the organisation by mean of the implication of the users in the implementation phase. It is shown that the implementation method may generate harder constraints than those coming from the system itself. Also, the explicitation of implicit transversal processes seems to allow an increased control of the individual's work, through standardisation of the spaces of confrontation. Information is extracted from its context and is supposed to be meaningful on its own; its standardisation according to global issues can be contradictory with local process, dealing with situated action, the necessary answer to unpredictable events, and context specificities. A question can be whether a relative local inefficiency is not considered as acceptable in comparison with a better global efficiency.

These are the main results of a three-year research programme concerning the relationship between organisational change and information system transformation in firms dealing with ERP implementation projects. This research programme has been funded by the CNRS, and more precisely by the "Information Society" Research Program (Bazet et al., 2003; Bazet and Mayère, 2004, 2007). Through three main case studies, our enquiries have mostly concerned the implementation process and its first results in the firms activities. Interviews have been conducted with members of project teams and expert users working with the implemented ERP (25 interviews), and with consultants specialised in ERP consulting and implementation (10 hours of recorded interviews with specialist consultants).

4.2 ERP and Information Production Design

4.2.1 The "Scientific Organisation of Labour" Applied to "Ordinary" Information and Knowledge

ERP are based on databases which are shared by the different functions and units of a firm: product databases, client and provider databases, as well as databases for all the resources required by the activities, including human resources. So as to be part of such databases, information has to be standardised. It has to be codified according to a format which is often established in its final form at the firm level, the global level. This happens to be often fairly different from what happened previously, when various codes were used in the different units or entities, or according to the professionals: designers, or manufacturers, or sellers, or technicians in charge of the technical support, often had codified differently information, as they were concerned with different dimensions or points of view regarding this information.

Not only is the information standardised, but also the information production itself. The standardisation logic is applied to the treatments which are carried out so as to identify information, and to process it, all along the linked activities. For example, a client order will go through several stages, and this will be specified.

This specification is usually done in relation with "expert users" who collaborate with the project team, which often include computer programmers. The information production "model" described in such procedures will form the references for the ERP configuration, specifying what will become the necessary way of processing the concerned tasks.

This evolution contributes to a stronger industrialisation of tertiary activities in firms, and more precisely of activities with intellectual dimensions, that were previously either not, or at an individual level, computerised. The "scientific organisation of labour" that Taylor promoted at the turn of the last century is therefore extended to information activities, and partly to intellectual tasks and skills (Bazet and de Terssac, 2007).

4.2.2 A Focus on Information which Can Be Formalised, and on Basic Exchange of Information

As mentioned earlier, "expert users" are asked to specify the information required as an "input" for the tasks they operate as part of their job, and to describe the treatment they apply to this information before sending it to the following stages. The focus concerns the treatments by themselves. The analysis does not involve the knowledge and know-how required to identify the information, its meaning and usefulness for the on-hand activities, and to coordinate with other employees and departments for setting up a common point of view on the questions to be solved (Grabot, 2007). In this respect, it is a quite partial view of the whole information and communication tasks and knowledge required for carrying out the activities.

We observed more specifically an ERP implementation in the purchasing department of an internationalised firm. We noticed that, according to the ERP implementation process, it was considered relevant to pull apart the required information production from the negotiation with suppliers. But these two processes are strongly related in purchaser everyday life. A supplier may be reluctant to negotiate a new order if the previous ones have not been paid, even more so if the purchaser does not know why and has no access to the answer, because of the division of information production labour formalised through the ERP configuration.

With digitisation and standardisation associated to ERP, quantitative and factual data take precedence over more qualitative and variable ones. This precedence was fully visible during the configuration process through our case studies. This configuration process relied on the questioning of expert users concerning their possible "needs" or requirements. The technicians in charge of this identification had to follow a highly formalised questionnaire, with obligatory criteria to be met to identify a requirement: it had to be measurable, short term, specific, accessible, and allow follow-up.

Such a definition of a requirement circumscribes the elements and situations that have to be taken into account. There is no room for ambiguity, for vagueness, which play, however, a fairly strong role in organisational problem solving according to James March analyses (March, 1991; Mayère and Vacher, 2005). All the situations are supposed to be predictable. However, contemporary firms have to be flexible, to be efficient with decreasing resources, they have to deal with growing inter-dependency with their suppliers through just-in-time purchasing, all circumstances that tend to sustain the risk for unpredictable events, and the need for ad hoc resolution.

All along the specification process, the metaphor of a flow is applied to information production. According to such an approach, it is supposed that communication is unnecessary, that a co-construction of the meaning of information is not required. The meaning of information and its possible use are supposed to be totally formalised in the information system (Levitan, 1982). An implicit statement in such an approach is that all that is necessary for conducting these tasks can be specified. This is not far from the scientific organisation of labour: each task is designed in such a way that it fits with the following ones, and communication is unnecessary. If employees try to communicate with each other, it may be considered as hanging around.

Such an approach of information and communication standardisation is questionable in firms dominated by flexibility, adaptability, whose resources are maintained at their minimum level. In such firms, there is a great need for specifying what are the current priorities, what problems are to be solved. This recurrent identification of priorities is aimed at answering the different client requirements, taking into account the constraints both inside and outside the firm.

In the observed purchasing department, the employees know that the stock in hand is very low. They are concerned with the delivery time of the suppliers, and with the production planning constraints. A great variety of events can raise obstacles to their main objective, that is to say: have the right supply delivered in good time. Before the ERP implementation, they used to deal with such contradictory constraints through flexibility in the information production process to allow adaptability in the production process: for instance, an agreement by phone, confirmed afterwards by the formal procedure, that was possibly completed after the supply had been delivered. Facing events in such a way relied in great part on personal trust and commitment. This is also a dimension of the work that is at least partly out of the control of managers.

In this firm as in other firms studied, we observed that ERP implementation was considered by the managers as an occasion for reasserting the formal rules. In this respect, ERP technology was used as an argument to make employees apply the rules according to the formal organisational design, breaking down the socio-political barriers, or trying to do so (Boitier, 2004). Through ERP configuration, these rules are formalised and therefore imperative: what each employee should do, what he or she could not do anymore, is defined according to his or her password protected access. The so-called "protection" is a two-fold one, including the guarantee for the managers that employees will stay within the borders of their formal role. In such a renewed context, employees have to play by the rules, even if this could imply not being able to fit with their activity objective. This is a typically "double bind" situation: employees cannot succeed in carrying out their main activity, which forms the basis of their performance appraisal, because they have to match information production to ERP configuration. These contradictory logics prevent them from doing their job; this is typical of what occupational psychologists point out as one of the contemporary main sources of occupational disease (Dejours, 1998). We observed that employees, faced with such a dilemma, were making attempts to change their access to the information system at a local scale, trying to get round this renewed set of rules attached to the use of information system.

4.2.3 Logic Priorities and Questions of Sense-making

When the priority is given to global reporting efficiency, a decreased efficiency can result at a local level from the configuration choice. In the purchasing department case study, the process analysis was composed of three main periods. It began with a first period dedicated to the local level; expert users in the units were asked to discuss together the way they were carrying out their activity. During the second period, a similar discussion was conducted at the regional level, that is to say, Europe, North America and Asia; negotiations took place to find a common process. The third period was dedicated to discussion at the global level; it took place in the US headquarters, with a great majority of US experts.

At this stage, the top management imposed a "no option" rule. The selected model for activity configuration has therefore been the American one, ignoring the arguments concerning the market regional differences.

Another debate took place concerning the language: will the regional components have the opportunity to choose it according to their culture, using a facility offered by the ERP software, or will it be a single one for the whole firm, namely American English? The debate had just begun when the top management imposed this second choice, for global reporting efficiency. The possible obstacles to sense-making at the operational level were not considered as a strong enough argument.

4.2.4 Inter-changeability of Information Producers

The arguments mobilised to justify ERP investment usually deal with upgrading productivity, reducing the costs dedicated to maintaining software and local databases, while increasing the speed and quality of reporting to headquarters. However, there is at least another goal, although rarely mentioned, which appears to be fairly important. That is the codification of information and of the information process which are required for all basic information activities in the firm, and their computerisation.

This dynamic contributes to a new type of "capitalisation of knowledge". The perimeter of such a knowledge capitalisation is quite different from what was meant by such a term in the 1980s, but it tends to be a very important stake in such a flexible firm type. When human resources have to be as flexible as other resources, the computerised information system tends to be the unique stable component of the overall information system, which includes informal communication and individual or local information production. The substitution between employees has to be easy for the smooth running of the business. Firms have to make sure of the inter-changeability of information producers, considered as part of information sources and of information treatment providers.

This substitutability also concerns organisation components. In contemporary firms, the top management has to plan what will be the firm perimeter in the coming years. This includes the choice of the activities, and units that are considered as being part of the business core, and the ones that are not, and could be externalised. The rationalisation of information production and of communication makes such externalization easier – or at least is supposed to do so – through the setting up of an organisational design which is independent of the agents or organisational components. There is, however, a fairly strong and risky underlying hypothesis: that the formal and standardised process is sufficient to carry out the activity, to face up to the complex and highly variable situations.

4.3 ERP Combined with Business Process Re-engineering and Business Process Outsourcing: Re-designing the Organisation While Transforming its Information System

4.3.1 Value-adding Versus Non-value-adding Activities

In order to specify what should be in the scope of the ERP, and what should rely on specific software, ERP consultancy service experts draw a line between activities which are supposed to provide the added value of the firm, and other activities. According to these experts, the first ones should be computerised with the support of specific software, so as to capitalise in the firm its core knowledge and skills, its main know-how, know-whom and whatever knowledge which makes the difference with competitors. The ERP package is generally concerned with the second type of activity, which is not specific to the firm.

The "main stream" point of view has changed regarding this issue during recent years. Previously, the main ERP editors were putting forward the "best practices" design approach so as to argue that all the information system should be integrated under the same ERP.

The ERP hegemony has thus been criticised and the interest of developing links with specific software is recognised, in relation to the issue of maintaining the competitive advantages of the firm.

However, one main question remains on the very possibility of identifying such differences between activities. How do we make sure that certain activities provide the added value, and others are just ordinary ones, without specific added value? According to consultants, managers do not always have a clear idea of what makes their efficiency and competitiveness. But consultants often do not have expertise in their client firm domain: they first of all are experts in ERP itself. Our observations show that their judgment concerning value-adding and not-value-adding activities reflects the shared ideas in the firm and its direct environments: it is often a mirror of usual thinking rather than an effective expertise. However, it may have very strong impact on firm re-engineering.

Suppose experts can manage such an identification. Even then, a lack of coherence may appear between the ERP standardised tasks and the other tasks. Business is made up of strongly interlinked routines and more ad-hoc activities. The fact of dividing them between two different sub-systems can develop contradictory factors which make the overall process inefficient.

The "thin" firm, based on "lean production" is sometimes not that far from anorexia, and the lack of redundancy can be quite risky in a just-in-time organisation. What we observed in the purchasing department tends to underline the links between very usual information treatments, and the core tasks of employees, namely, purchases. The limits of the optimised process appear when nobody knows when such invoice will be paid, when nobody can tell what is the possible obstacle, and how concerned people could get round it.

4.3.2 Re-engineering, Outsourcing, and the Robustness of Information Processes

ERP implementation is often linked to re-engineering and outsourcing. Once optimised, the information production process relies on a division of labour which combines internal and external resources, and units in different countries. The employees processing information at a certain stage do not necessarily know who will do the next treatment, who did the previous one, neither when or where. Part of the knowledge required for the information production and for the overall business can possibly be capitalised in the ERP databases and procedures. Is it sufficient for carrying out the business, particularly in the current flexible economic environment? With such a division of information production labour, is the intelligence required for the activities still somewhere in the firm? One can express some doubts about it when observing at a detailed level the way this evolution is going on in firms.

4.4 Contradictory Dynamics Relying on Local Employees and on Project Teams

4.4.1 Project Teams Dealing with Local Versus Global Contradictory Dynamics

The project teams in charge of the ERP configuration have to deal with complex dynamics which include contradictory components. They have to obtain the active participation of expert users so as to clarify the existing procedures and process. They also have to impose on the so-called "final users" the standardised codes when confirmed by the management (global level). In internationalized firms, the local specifications are often discussed and finalised by the headquarter managers according to global priorities – namely, optimising the reporting – and this can imply important changes in the formats defined at the local stage. When such a global standard is sent back to the units, project teams have to accompany final users in the appropriation process, when they try to understand and use the renewed procedures (Orlikowski, 1992, 2000).

Giddens has pointed out the delocalisation–relocalisation process as one of the main dimensions of modernity (Giddens, 1994). The contradictory dynamics that project teams have to deal with can be analysed within such an analytical framework.

4.4.2 The Selection of "Expert Users"

First of all, the question of who are the so-called "expert users" has to be specified. In one of our in-depth field studies, the expert users were the persons in charge of the departments which were meant to be concerned with the ERP project. These managers had no direct experience of the activity, and they were relying on their guess concerning the way the activity was or should be carried out. The risk of a gap between their perception of the activity as the "usual way it should go on", and its variety as a fact, was in such case fairly high.

Usually, expert users are employees who have long experience in the concerned activity. This could be questioned also, taking into account the computerisation logic combined with the flexibility of human resources and the constraint of employee substitutability. Employees who have carried out the activities for years master a whole set of implicit knowledge which is useful for identifying the meaning of information. They are able to distinguish the possible differences between current indicators or data and usual trend of activities. This may be not the case for newcomers, who will have the opportunity of relying only on the computerised process, without additional expertise for assessing the on-hand activities.

In another case, the expert users happen to be fairly recent ones. In the purchase department of one of the studied firms (Electronic), there was previously an important change which resulted in the fact that all the employees of the department were newcomers when the ERP project was submitted. This was not taken into account when deciding who would compose the expert user team. These newly arrived employees were asked to describe what the procedures were.

Although the expert user role appears to be a very important one, it seems that pragmatism comes first when deciding who will be part of this process, ignoring the risk linked to either too much or not enough expertise.

4.4.3 The "Expert Users": the Gap Between First Hopes and Final Results

Our observations, as well as other research programmes come to the same conclusion: the "expert users" collaborate quite actively to the codification process (Grabot, 2007). Managers present the investment in ERP as one of the "naturalised constraints" of the contemporary firm. ERP investment is talked about as an evolution that could not not happen (Durand, 2004). Therefore, employees focus on "how to" rather than enquire "why". Heavy constraints imposed on planning, with short delays and strong pressures for keeping in the framework, are participating to this viewpoint (Thine, 2007).

The ERP implementation goes through different steps: one deals with the process analysis. It is meant to identify the different tasks which take place all along the process under study. During this stage, the employees are asked to speak freely on the ways they perform an activity and on the possible improvements. The project team is usually composed of employees working in different parts of the firm, and the persons involved are often quite satisfied with such an opportunity to discuss what they do, and knowing better what other employees accomplish in other departments or functions.

A further step usually deals with knowing better what the selected ERP will require for process specifications; and the following one is dedicated to make the process analysis converge on the ERP requirements. During these two stages, the expert users often do not play an important role, and may even be totally absent. Consultants often play the main part. They have a propensity to rely on their previous experience of the ERP to specify the computerised process in the current firm. Their investment in the firm meets heavy constraints, because of the high fees and the often short time that is funded for this project. What often occurs, according to different observations, is an important gap between the process analysis as specified by the expert users, and what is finally presented to them as the computerised process according to the ERP constraints (or, but this is not said, according to the consultant's knowledge and the means devoted to ERP specification).

This gap engenders fairly strong disappointments within the expert users, and through them, within the final users (Grabot, 2007). The feeling is that all these efforts have no result, that the complexity of work is denied, or even that the process analysis was only a way of getting employee involvement before organising convergence to an already defined goal.

Expert users, when involved at this stage looking for a tight fit between the process on hand and the ERP constraints, are mobilising rudiments which are part of knowledge and competences associated with their job. Facing such a supposed-to-be unavoidable change, they try to make sure that they will still be able to carry out these tasks after ERP implementation. Short term issues tend to predominate over the risk linked to knowledge capitalisation at the firm level, in a context where nobody really knows what will be his (her) job and employer in a few years or even months. However, as mentioned earlier, the capitalisation process at the firm level tends to be limited to the basic data and treatment of such data, as long as the knowledge and know-how required for transforming data to information useful for action is not involved in the process.

4.5 Back to the Definition of Information and Knowledge Associated with ERP Design

4.5.1 Knowledge Management Renewed Through ERP Projects

ERP projects aim at covering most of the functions and departments of the firm. Thus, they fulfil a hope held since the 1970s by information systems specialists: to get rid of specialised and isolated information sub-systems which were specific to departments, units or functions, and to set up a unified information system.

However, what is at stake does not only concern the setting up of a single and integrated information system. One main issue is the extensive development of information computerisation. For instance, in sales departments, sales representatives can be asked to record the characteristics of their clients, the needs expressed and the questions submitted, that is to say, very specific information which was previously stored in personal minds or on personal registers. Such dynamics sustain the formalisation of different types of information, from the most quantitative to more qualitative ones. This formalisation, combined with computerisation, transforms personal information into shared information at an organisational level. Each employee is asked to produce information not only for him or herself, but for other employees, most of whom they never meet and do not know. This information production assumes a growing part in the overall tasks of employees (Grabot, 2007). However, it is often not taken into account in job description. On the other hand, employees often share a depressive point of view on information production, which is linked to boring administrative constraints, or to low level jobs (often combined with gender discrimination). This point of view can be enforced by the growing part of information production which has no local use, and thus can be looked on as unnecessary.

The formalisation of information began long before ERP projects, but such projects sustain this evolution very steadily. Through ERP projects, firms are setting up a very basic and pragmatic form of knowledge capitalisation. It is a rough type of KM, compared with the highly sophisticated dedicated information systems that were developed during the 1980s. This current "ERP type" of KM could be in some ways more efficient than the previous ones, because it is widespread and because it formalises the quite usual information required for carrying out the functional tasks in the firm. However, its weakness results from the fact that it does not include the knowledge and know-how required so as to transform data into information, that is to say, meaningful basics which help understand the complexity of on hand activity, and specifying the required decision. We will develop this argument further on; it concerns the misunderstanding of differences between data and information which is at stake in ERP implementation process methods.

4.5.2 The Need for Going Back to Definition: Data, Information and Knowledge

ERP specialists usually refer to "data", which are considered as the basic unit required for the smooth running of ERP. Information in such approach is considered to be equivalent to data. It is considered as a "raw material". It is supposed to include all the dimensions required for its direct use. However, information and communication sciences have pointed out the fact that data have to be associated with existing knowledge in order to get some meaning, that is to say, become information. Existing knowledge is contextualised. To be meaningful, information has also to be linked to on-hand activity, be it material or intellectual. Thus, information is temporally and spatially situated. Data form a potential basis for information; they are not information by themselves.

Computer scientists have until recently often neglected the information and knowledge characteristics from the user's point of view as they tend to recognize it themselves nowadays. The problem is even worse concerning ERP logics, because it is combined with the specific point of view associated with accounting uses.

ERP software has first of all been designed according to the model of accounting information. Accounting information has to be formalised in single data, which are supposed not to be modified as soon as they have been officially registered. These characteristics are very specific to such an information domain.

The reporting to head-quarters by the firm is often the main purpose of the ERP investment. To facilitate such reporting, the priority is given to quantitative data. It is assumed that all information can be formalised according to accounting information rules. ERP design logic relies on the hypothesis that the usual information mobilised to carry out the activity has the same characteristics as accounting information.

But in "real firms", a great deal of information is recurrently transformed, reconsidered, adjusted according to the activity itself and to the environment evolution.

4.6 Conclusion

In conclusion, ERP design underestimates the situated knowledge required so as to transform data into useful information. From this point of view, it is still fairly far from a KM information system. But managers may think that it is similar to a KM information system, which is a fairly risky thought. They can guess that the flexibility of human resources can be managed through databases and standardised procedures, missing the fact that this does not produce the intelligence required to pilot complex systems.

The second main result is that there is disruption between the data in ERP software designed towards reporting aims, and the production of information and knowledge required for carrying out the activity. The very question of the meaning and possible use of information in a situated action is avoided, because global logic (reporting) predominates over the local one (dealing with the activity under current constraints). What appears through our enquiries is that employees in charge of activities try to handle this gap. They try to deal with the various formal and informal information systems in order to make out, within existing information, useful meaning for solving the various problems that are induced by modern "lean organisation", just-in-time production and customised product requirements.

There is some doubt that employees will be able to deal with this gap for long. Wage agreements have been weakened by successive re-engineering processes and they do not include long term commitment. The employee involvement in the firm is destabilised in such a social context.

Some researchers insist on the fact that ERP principles were specified in the 1970s. They assert that ERP incorporate a certain idea of organisation and management as stated at that time (Gilbert and Leclair, 2004). Since then, firms have met tremendous organisational changes, and there could exist a great gap between ERP principles, and the firms they are supposed to fit. Our findings, which are confirmed by other research program results (Boitier, 2004), tend to underline the importance of the way ERP are put into service in the concerned firms. The way it is carried out may even have a stronger impact than the characteristics of ERP themselves. The technology is an argument for organisational change more than the factor of such change, and the informal part of technology, made of configuration methods, appears to introduce tighter constraints than the more formal computerised one.

A linked issue deals with the timing of such organisational change. ERP projects are often presented as long term ones. In ERP advertising, the firm is often presented as an entity detached from space and time. But the current firms meet important changes within a short while. They can be split in different parts, or combined to other ones. Then, the previous choices regarding ERP specification may be totally or partly denied by recent events. This has been observed in each of the firms under study. The question at stake concerns the resulting precariousness of information systems, and of the project teams in charge of ERP implementation.

4.7 References

Bazet I, Erschler J, Mayere A, De Terssac G, (2003) Construction sociale de l'efficacité et processus de conception – appropriation: le cas de l'implantation de PGI. Rapport final "Société de l'information" du CNRS, March

Bazet I, De Terssac G, (2007) La rationalisation dans les entreprises par les technologies coopératives. G. De Terssac, I. Bazet, L. Rapp (Ed.), La rationalisation dans les entreprises par les technologies coopératives, Ed. Octarès:7–27

Bazet I, Mayere A, (2004) Entre performance gestionnaire et performance industrielle, le déploiement d'un ERP. Sciences de la Société 61:107–121

Bazet I, Mayere A, (2007) Ingénierie organisationnelle et réification des informations: le cas du déploiement des ERP. G. De Terssac, I. Bazet, L. Rapp (Ed.), La rationalisation dans les entreprises par les technologies coopératives, Ed. Octarès:81–94

Boitier M, (2004) Les ERP. Un outil au service du contrôle des entreprises?. Sciences de la Société 61:91–105

Dejours Ch, (1998) Souffrance en France. La banalisation de l'injustice sociale. E. du Seuil.

Durand JP, (2004) La chaîne invisible. Travailler aujourd'hui: flux tendu et servitude volontaire. Ed. Du Seuil

Giddens A, (1994) Les conséquences de la modernité. Paris, L'Harmattan.

Gilbert P, Leclair P, (2004) Les systèmes de gestion intégrés. Une modernité en trompe-l'œil ?. Sciences de la Société 61:17 – 30

Grabot B, (2007) Vers une meilleure prise en compte des utilisateurs lors de l'implantation des ERP. G. De Terssac, I. Bazet, L. Rapp (Ed.), La rationalisation dans les entreprises par les technologies coopératives, Ed. Octarès:125–147

Levitan K, (1982) Information resources as "goods" in the life cycle of information production. Journal of American Society for Information Science 33(11):44–54

March JG, (1991) Décision et organisation. Dunod

Mayere A, Vacher B, (2005) Le slack, la litote et le sacré. Revue Française de Gestion, hors série Dialogues avec James March:63–86

Orlikowski W, (1992) The duality of technology: rethinking the concept of technology in organizations. Organization Science 3(3):398–427.

Orlikowski W, (2000) Using technology and constituting structures: a practice lens for studying technology in organizations. Organization Science 11(4): 404–428

Thine S, (2007) Redistribution des rôles dans l'urgence du déploiement de l'ERP. G. De Terssac, I. Bazet, L. Rapp (Ed.), La rationalisation dans les entreprises par les technologies coopératives, Ed. Octarès:95–105

5

Why ERPs Disappoint: the Importance of Getting the Organisational Text Right

James R. Taylor[1], Sandrine Virgili[2]
[1]University of Montréal
[2]University Paul Verlaine Metz, CEREFIGE

"One survey of ERP project managers found that 40% of respondents failed to reach their original business case... more than 20% of managers stated that they actually shut down their projects before completion." ERP projects were "being delivered late and over budget with costs that were on average 25% over their original budgeted amount." Firms "have spent on average $48 million to date on ERP projects that are only 61% complete." – Beatty and Williams, Communications of the ACM (2006).

5.1 Introduction

In 1994, the journal Computer Supported Cooperative Work (CSCW) featured a debate between two acknowledged "stars" in the field: Lucy Suchman, then a researcher at PARC (the Xerox Palo Alto Research Centre), with a background in ethnography, and Terry Winograd, a professor of computer science at Stanford University, also located in Palo Alto, California. Suchman led off the polemic with an article entitled "Do categories have politics?" In her paper, she opened up for argument the validity of all computer-based systems that claim to be "tools for the coordination of social action" (p. 177). She questioned in particular how "the theories informing such systems conceptualise the structuring of everyday conversation and the dynamics of organisational interaction over time" (p. 178). Her explicit target was a system, called "The Coordinator," that Winograd had been instrumental in developing. It based its protocols on a theory of organisational communication derived from earlier work in philosophy, linguistics and discourse analysis known as "speech act theory" (SAT). SAT proposed a categorisation of utterances, based on how they contribute to a set of presumed standard organisational transactions, which The Coordinator proposed to make explicit and incorporate as part of a computer-based protocol. In this way, it claimed, the supporting technology it offered would render the communicative exchanges of the organisation more transparent, and thus—by implication—would make them increasingly regular and efficient.

The Coordinator was, in fact, one of many predecessors to today's much more commercially successful ERP technology: one of the dead ends along the path. It was motivated by the perception that organisation may be thought of as an assembly of transactions that collectively add up to an internal economy. Such transactions are normally accomplished in an ongoing universe of conversations, where individuals and groups negotiate the arrangements that enable them to coordinate the timing and terms of their collaborative efforts. Mostly, this has traditionally been part of the informal background talk that people use to smooth out their efforts at cooperation. The Coordinator promised to render these conversational exchanges more transparent. It was, as noted above, inspired by speech act theory whose originators, John Austin at Oxford and John Searle at Berkeley, had proposed a categorisation of acts of speech that amounted, in the hands of linguists, to a claim to have identified the underlying syntactic/semantic underpinning of human interaction. The Coordinator thus aimed to formalise and standardise the informal background conversation typical of all organisations.

Suchman's critique focused on the crucial assumption that "explicitly identified speech acts are clear, unambiguous, and preferred" (p. 180). Sometimes, it must be admitted, a question really is a question, and a request really is a request. At other times, however, a question is actually a request, and a request is in fact an order. Knowing which is the "real" meaning, what is explicitly said versus what is indirectly implied, is something people do quite well, and language-based machines not as well. Suchman therefore doubted the claim of SAT that the intention of any act of speech "is somehow there already in the utterance and that what is being done is simply to express it" (p. 180). The meaning of an utterance in real conversations, she countered, is open-ended and negotiable (there is indeed an impressive body of empirical evidence to back her up on this score, drawn from a field known as "Conversation Analysis" or CA). A measure of ambiguity, CA researchers have documented, is inevitable in any real interaction. And, more important, what if, as Eisenberg (2007) has argued, ambiguity is not an index of sloppy language use, or inefficiency, but an indispensable cushion that renders organisational processes effective—a crucial lubricating oil that prevents relationships from deteriorating into open opposition (Goffman, 1959)? And, if that assumption is valid, why would you want to eliminate a vital contributor to the frictionless operation of the enterprise: what ethnomethodologists call the indexicality of language-in-use, namely its dependence on context and circumstance for the decoding of its meaning?

For Suchman, the introduction of standardised protocols of interaction thus had less to do with clarity of purpose, or efficiency, than with discipline: to create "a record that can subsequently be invoked by organisation members in calling each others' actions to account" (p. 181). Citing Foucault, she accused The Coordinator's developers of complicity in the veiled exercise of power. "For management," she wrote, "the machine promises to tame and domesticate, to render rational and controllable the densely structured, heterogeneous nature of organisational life" (p. 185). It would become "a tool for the reproduction of an established social order" (p. 186). The computer scientist, she went on, "is now cast into the role of designer not only of technical systems but of organisations themselves" (pp. 186–187).

In his reply, Winograd in turn accused Suchman of blatant over-dramatisation. He poured scorn on her attribution of a sinister motive behind the development of the system. "The sub-text," he wrote, "is a political drama, in which the villains (corporate managers and their accomplices: organisational development consultants and computer scientists) attempt to impose their designs on the innocent victims (the workers whom the managers want to "tame and domesticate")" (p. 191). As against this Faustian tale of the clash of cosmic forces of oppression and liberation, Winograd offered a more mundane account. As he observed, "one could take the contrary view—that the regularity provided by explicit categories and disciplines of bookkeeping makes possible whole realms of collaborative production of social action that would not exist without a regularised structure that is mutually understood and obeyed" (p. 194).

To buttress this less emotionally charged (if equally contentious) interpretation, he cited the homely example of his own grandfather who, earlier in the century, had started a small business. As long as it stayed modest and local, he could run the whole operation out of his hip pocket, with the accounting kept mostly in his own head. But when the enterprise began to grow, with more employees, he had to introduce systematic bookkeeping. Apart from any other consideration, the Internal Revenue Service expected something more reliable than one individual's memory; they wanted to see "the books." You cannot, Winograd pointed out, "run even a moderately small company," much less a company with 10,000 employees and thousands of suppliers, "without regularised (disciplined) accounting procedures" (p. 194). "Imagine," he went on, "a world in which every business invented its own accounting procedures, or in which each person in an office adapted them in arbitrary ways" (p. 194). The result, he concluded, would be to "create unbearable chaos in all of those areas where people needed to interact" (p. 194). Agreed, he wrote, any organisation is a "web of conversations and commitments among the people inside and outside the organisation" (p. 194). But they have to be kept track of in a disciplined way, if the company is to work at all.

(Of course, Winograd's argument does rest on the implicit assumption that the categories of the computer-supported system are, to use his phrase, "mutually understood." That, it turns out, is also the problematical component of an ERP implementation, as we shall show later in the chapter.)

The debate, in one respect, can be interpreted as an encounter of contradictory conceptualisations of the relationship of an organisation to its members. Two contrasting images of the basis of organisation lie behind the respective positions—two metaphors (Morgan, 2006). In Winograd's image, organisation is a rational configuring of interlocking activities to produce a coherent collective actor, capable of growth. For Suchman, organisation is a dense web of work and talk that develops its own internal coherence, and modes of being. In consequence, Suchman's argument (as the title of her piece suggests) came down to the issue of categories, and, even more important, whose categories are the more important— those of the productive working majority, or those of a privileged few, isolated at the top, aided and abetted by their professional advisors. Winograd, for his part, retreated to safer ground, where the debate is interpreted differently: whose interest should take precedence, that of the organisation or those of its members. Indeed, are they not in the end the same, he implied?

This difference of perspective reflects, of course, one of the enduring puzzles of organisational theory, and is unlikely to be soon resolved. The authors of this present chapter, however, see this debate somewhat differently. For us, the systemic-humanist polemic is ultimately grounded in one of the great philosophical debates of the twentieth century, personified by Ludwig Wittgenstein. Winograd's position has its historical roots in a conceptualisation of communication (and language) as a vehicle for the conveyance of information, and the exchange of knowledge. This is a theory of communication whose rationale can be traced back, in part, to the dramatic advances in the formalisation of logic that dates from the late 19th, early 20th century work of Boole, Frege, Russell, Hilbert, Gödel, Turing, von Neumann—as well as the earlier Wittgenstein. It is founded on the assumption that language is, above all, a tool for the formulation of our understanding of the world into an equivalent representation, expressed in the strings of symbols, or "formulas", that we usually think of as sentences. The business of logic, they reasoned, would be to discover the fundamental underlying structures of meaning that often become blurred in the more complex syntactic/semantic hybrids of actual speech—somewhat like the designers of The Coordinator hoped to make the transactional dynamic more transparent and regular. If the logicians could isolate the essential core of meaning then it would furnish the most transparent possible instrument for conveying knowledge.

The invention of the computer, in this perspective, was merely an effective way to mechanise the core structures of meaning: make them socially useful in the sense of more productive. The development of a mathematical theory of communication by Shannon and Wiener (1949), in the late 1940s, simply expanded this tradition by establishing a reasoned technical basis for the efficient transmission of such logic-based kernels of meaning. ERPs are one current manifestation of this philosophy, and the practices it supports.

The problem was that by mid-century influential philosophers were questioning the basic premise of this whole movement: the notion of logic (including applied logic) as a linguistic vehicle for the statement and sharing of facts. The most striking of these reversals of perspective is exemplified by a rejection by the later Wittgenstein of the principles embodied in his earlier writings. His posthumous book, Philosophical Investigations (1958 [1953]), set out to debunk the entire logical positivist claim to neutral objectivity. In his preface Wittgenstein wrote: "I have been forced to recognise grave mistakes in what I wrote in that first book" (the reference is to the Tractatus Philosophicus, published in 1921). The essential "grave mistake" that mattered was the assumption that the business of language (or logic) is to record, and make generally available, "facts" about the world. Wittgenstein now proposed an alternative theory of communication, based on the principle that language is inherently tied to practice. It is about how people use language to do things. Because people use the same words to do different things, the expressions of language do not – cannot – have constant meanings across contexts, where such contexts differ significantly from each other in participant activities. Trying to fix the meaning of facts by recording them in a formal protocol such as computer-based accounting systems is a labour of Sisyphus—doomed to eternal frustration.

Computer scientists are, of course, hardly unaware of the difficulty of what Hoppenbrouwers (2003) has identified as the exigency facing all computer-based design: to "freeze language" (the sub-title of his dissertation was "conceptualising processes across ICT-supported organisations"). His study focused on a service agency in the Netherlands, responsible for social insurance and reintegrating unemployed workers back into active practice as soon as possible. His interviews unearthed the reality that the "same" operational term defined by official policy, and inscribed in the accounting system, was interpreted differently from one district to another. There was puzzlement as to the meaning of the official categories that, incidentally, formed the basis of the existing computer text. People, in the everyday circumstances of work, simply made up their own interpretation of provisions in the act that authorised their agency. The practice, naturally, varied from office to office. As Hoppenbrouwers noted, people felt alienated: "ICT people do not speak our language," they intimated to him (p. 202: ICT language, of course, originated in the "language" of logic as the younger Wittgenstein understood the term).

Hoppenbrouwers' intent as a designer, to "freeze" the language of categories, was not, he made clear, a refusal to take into account the importance of "the intuitive ability of people to use and interpret language flexibly" (p. 22). Instead, he simply aimed to narrow the gap between categorisation and actual usage from both ends: by making the official categories more comprehensible, and by taking account of actual practice in establishing them. Winograd made essentially the same argument: of course not everything can be reduced to computer code, but there is ample room for improvement in organisational performance overall, short of perfection.

The object of this chapter is to build on this and similar initiatives. We accept the validity of the respective points of view voiced by both Suchman and Winograd, in that we assume that there is no cut-and-dried solution to the paradox of organisation. It is, and must be, at one and the same time, integrated and differentiated (Lawrence and Lorsch, 1969), homogeneous in certain respects, heterogeneous in others, both formal and informal. Like Hoppenbrouwers, we seek, not a "solution," but a better understanding of the dynamic that the implementation of a new system such as an ERP triggers.

Our analysis draws on a field case study of an introduction of ERP technology into a large firm, one which exhibits the kind of compromises that must be made between modes of language use that illustrate the different ontogenies of organisation: system versus practice (Brown and Duguid, 2000). Our research is grounded in the contemporary theory of organisational communication, a perspective that sees organisation as an intersection of two modes of communicating, through conversation and through text (Taylor et al., 1996; Taylor and Van Every, 2000). It is in the turbulence generated by the mixing of modes that the origins of organisation are located, where "organisation" is conceived, not as a fixed structure, but as an organising (Weick, 1979). Organisation is the outcome of a hybrid enactment: both a formal system of laws and regulations, and an informal domain of open-ended and continuing sense-making. The implementation of an ERP, because it upsets established modes of organising, generates zones of what Weick calls "equivocality," and triggers cycles of sense making, in which more than practice is at stake; so are its rules. Identities, and patterns of authority, are also made problematical. When the stone is dislodged, the ants scurry to re-organise.

Our chapter is organised as follows: first, we develop a brief exploration of the theory of organisational communication; second, we present and comment relevant findings drawn from the case study; third, we conclude by some observations on the contradictory textual bases of technology and organisation.

5.2 What is an Organisation (and What Is Its Basis in Communication)?

The Suchman – Winograd "religious war" (de Michelis, 1995) stimulated a vigorous continuing debate on the issues they had raised, which was published in the same journal, CSCW, the following year, 1995. At the core of the issue for Suchman, as she now made clear in her response, was the question of "whose notions of organisational life" were being represented: those grounded in the "rationalities of technology design" (what we often tend to think of as the domain of text) or in the "actualities of use." As King (1995), in his contribution, observed, the debate was in fact "a replay of an ancient conflict over speech vs. writing" (p. 52), one whose origins he attributed to Plato, among others. King went on to observe, "Speech act theory makes sense only in the transparent realm of spoken discourse, wherein nuances of meaning can be sorted out and, by implication, sophisticated negotiation can occur. ... A performative speech is less about making promises than about making deals. Suchman's concern is that any device that "reduces" transparent speech activity to writing activity would, in use, severely compromise the establishment and leverage of shared meaning essential to the development of shared understanding" (pp. 51–52). Against this argument, King writes, Winograd cites "pragmatic necessity, not for The Coordinator per se, but for writing in general. Writing is necessary due to the inherent limitations of speech" (p. 53). Anyway, as King notes, he had claimed that "individuals using tools like The Coordinator can readily default to the domain of speech if the constraints of writing become too onerous and dysfunctional" (p. 53).

ERPs, fully as much as The Coordinator, must, by their very nature, "compromise the establishment and leverage of shared meaning." Yet the "compromise" cannot be avoided if the organisation is going to remain adaptive to its environment. The minute you transcend the boundaries of the here-and-now of a local conversation—the intimate world of interactive speech—then you have no alternative: you have to resort to writing even though, as King puts it, it risks "sundering the critical access path to thought and meaning" (p. 52).

This is why Suchman focused on categories. All language uses categories: nouns, verbs, adverbs, conjunctions. We automatically discriminate between tomatoes and tamales, birch trees and beech trees, eggshells and eggnogs. Suchman, however, would have been particularly sensitive to issues of categorisation since they had been dividing the social science community for a quarter century or so. The earlier explorations in formal logic to which we have already referred had impacted not only on the domain of computing. They had become, through the efforts of the so-called "logical positivists" (the "Vienna School"), the bible of researchers in the social sciences generally. The trademark of this dogma was the presumption of such investigators that is was they, as "scientists," and not their "subjects" or "respondents," who would choose the categories used in research. Approved theory would conform to the logical calculus of facts-induction-conclusions. Any deviation from this strict model would be merely "impressionistic" or, even worse, "literary"—scientifically unacceptable. In the 1960s, however, inspired by the work of such pioneers as Cicourel, Garfinkel, Goffman, Labov, Sacks and Schegloff, a counter-movement took shape, called ethnomethodology. It was grounded in the belief that everyone, not just the social scientists, is in the business of categorising—making sense of what is going on around them. There are no universally valid "categories." Categories arise, as Wittgenstein had earlier argued, in a practice, and reflect the exigencies of such a domain of focused activity. The "practice" of the social scientists (or, for that matter, the computer scientists) has no essentially privileged priority: it too is just one more way of making sense—whether for better or worse being an empirical issue. The proof of the pudding, after all, is in the eating.

An organisation, since it is an amalgamation of many practices, also has many domains of sense making, each endowed with its own categories, and supporting modes of interpretation of the environment it is involved in. Brown and Duguid (2000) report on the dysfunctional result (from management's viewpoint) of this differentiation of specialised knowledge bases: large firms such as Hewlett-Packard develop an extraordinary fund of diversified knowledge, but, paradoxically, the "knowledge the firm can hold on to, it can't use. And what it might use, it can't hold on to" (p. 150). It is not easy for people who have mastered different "language games" (Wittgenstein, 1958) to communicate with each other (Barley, 1996). It is much easier with others who use the same language they do, even if they are outside the boundaries of the organisation. As HP's president vocalised the dilemma, "if only HP knew what HP knows."

This is the problematic we address in this chapter: how technology affects the indispensable balance between a crucial spontaneous and local sense making, mediated by conversation, and the extensions of such practices in time and space that technologies (notably writing, even when it takes the form of computer code) seem to offer. How is the conversation translated into the text and, vice versa, the text into the conversation? Since the "answer" to this question, we have contended, is an empirical issue, our manner of exploring the impact of ERPs on organisations is through case studies. As Grudin and Grinter (1995) observed, in their contribution to the CSCW debate, when a new system is implemented in an established firm, with its own practices, "of course these activities will not just be "entered into" and "supported", they will be changed" (p. 56). Sometimes, to be sure, the authors observe, "disruption may not be bad." Sometimes practices should change. But sometimes change is not so positive, and may actually depress the performance of the "learning organisation."

What we will be delving into in this chapter is both the theory and the nitty-gritty of such "disruptions": how, in practice, they manifest themselves, and what they mean in a larger perspective. As Malone (1995) put it, "we need to learn the "art" of applying categories well" (p. 38).

5.3 The Case Study

The site of our case study was a large company, to which we give the fictitious name Labopharma, whose annual income amounted to some 160 million euros in the year 2000, with an annual growth rate of about 10%. Labopharma is the European leader in its own field, specialising in what is called "phytotherapy," or plant-based medicine. It began operations in 1980, was an instant success, and is now counted among the 100 most profitable French firms. In 1996, the company went public, and entered into a phase of rapid development. There were, however, problems. Perhaps the most salient of these was the need to modernise the entire accounting system. Like many such enterprises that grow like Topsy it had implemented a veritable Babel of incompatible information technologies, each specialising in its own domain, and weakly interconnected with other systems in the network, if they were not all mutually incompatible. There were thirteen different computer-based systems in operation, depending on the domain: finance (6 systems), production and purchasing (2 systems), warehousing (1 system), sales (4 systems).

Labopharma now found itself under intense pressure (from shareholders and regulators, among others) to consolidate its information/communication technologies (ICT) and to implement an infrastructure that would be capable of furnishing a more complete, transparent and up-to-date comprehensive account of its business operations. In 1998 it decided to bite the bullet. It first hired a consultant firm to counsel it on how to proceed. On the latter's advice, management decided to adopt an ERP system (ERP stands for *Enterprise Resource Planning*). Internal committees were established, and a request for proposals issued. The company, however, set stringent limits on the budget allocated to the venture. The choice, finally, in 1999, came down to two bidders, those who had submitted the lowest price estimates. On the advice of the in-house head of information services the choice went to a supplier with international connections. Shortly afterwards, however, the company encountered financial problems, and withdrew from all of its operations in France, including Labopharma. The usual messy court case followed. But Labopharma still had no integrated system. In 2001, a new request for proposals was issued, and now Labopharma elected to go with the international leader in ERP technology, SAP, a German firm. A contract was signed later that spring.

The constraint, this time around, was an urgent need to implement the system in the shortest possible time. SAP reckoned it could meet the requirement, and fixed a target date of August 2002 for full operation of the new system, little more than a year later. But, to do so, it established some very exacting conditions. There would be, for example, no preliminary phase of needs analysis and tailor-made design to take account of the special character of the firm, and its established modes of operation, other than the one that Labopharma had already conducted, in collaboration with its initial contractor. Labopharma would be buying a ready-made, off-the-shelf system, one that SAP argued would suit its needs because it incorporated and exemplified the "best practices" of the pharmaceutical industry as a whole. The "solution," in other words, would dictate the definition of the problem, not merely for technical reasons but to ensure overall coherence. Where there were incompatibilities between current modes of accounting and those dictated by SAP technology, it would be the latter that would be given priority. There would have to be some adaptations, of course, but they would be minor, merely enough to assure rapid implementation and efficient operationalisation.

SAP, to meet this requirement for a shortened time horizon, resorted to a protocol of development known as RITS, or *Rapid Implementation Tools and Services*. A strict timetable was set: Phase 1, June – July 2001, initial planning and resource mobilisation; Phase 2, September – October 2001, identification of gaps between system and current practice; Phase 3, November 2001 – March 2002, adaptations necessitated by the gaps, development of interfaces, start of testing; Phase 4, January – June, data transfer, training; Phase 5, July – August, documentation, additional training and launch. The underlying principle?: "Big Bang." It would be, in other words, an overnight switch from the old systems to the new-computerisation on the run, as opposed to incrementalism.

SAP, through its consultants, began preliminary work on the project in the summer of 2001 (June-July), including detailed planning, assembling of resources, all conducted with the collaboration of Labopharma, but managed in-house by SAP's designated consultants. Basically, the work at this juncture consisted of a re-analysis of the planning the company had engaged in during the earlier aborted project. In addition, there were a host of details to be worked through: where meetings would be held, how to plan the intervention of the consultants who would manage the actual implementation, discussions of strategy with senior representatives of management. The actual launch did not take place until the months of September and October (it was in September that our own participation in the project began, from the very outset of the implementation phase). Only now were the operational company officers delegated to the project actually briefed on the details of the new system. What they discovered, as the project began to unfold, was disconcerting.

First, a word about the organisation of the working groups. Two committees were struck. The first was a steering committee, led by the senior management group, with representation from eight sub-project company heads, covering commercial and marketing operations, finance, production and administration, plus two implementation chiefs, one from the company and one from the consultant, aided by a change manager. This steering committee would meet as needed. At the level of the actual project, two categories of specialist were distinguished: in addition to the implementation chiefs, there were the eight sub-project heads already mentioned, and six computer specialists from the firm itself, again identified with the areas of commercial/marketing, finance, production and administration. This more operationally focused project committee would meet weekly. It included the project chiefs from the consultant and the company, a coordinator of the various information systems already in operation, plus the sub-heads. The committees were meant to smooth the transition, by identifying and resolving problems as they might arise. Where they did find issues, the various teams were instructed to submit a work report on any technical incompatibilities, specifying the nature of the gap between the expectations of the designers and consultants, and those of the company officers who had a more detailed knowledge of existing local practices. The actual work would, it was thought, usually be done by small working groups varying between four and six persons, seldom more than eight.

The procedure, to be more precise, consisted, first, in trying to visualise, for a given kind of transaction, the path it customarily followed, its connections with other functions, and the hierarchical organisation it necessitated (what authorisations it called for, for example). In some respects, the envisaged procedure was reminiscent of that of an archaeologist, tracing the indistinct lines of a long-lost city, to imagine the pattern of activities that must once have gone on there. As these usually taken-for-granted modes of operation were identified, and made more transparent, it then became possible to conceptualise the gap between current modes of working, and those that SAP envisioned. As this process transpired, however, the complexity of the SAP technology was also beginning to reveal itself. How to reconcile accepted practice and new system now became less a simple matter of identifying discrepancies and correcting them than it did of finding a way to deal with the intractable realities of practice either by modifying the technology, or abandoning the practice—or both. This was not exactly the way the development process had been envisioned. It was more complex—*considerably more.*

Let us consider one example of what we are referring to: managing shipping operations. The technology SAP envisioned worked on the basis of individual orders from a client, line by line; the usual practice, however, was predicated on dealing globally with an overall order. Here is how one sub-project head, interviewed informally, explained the problem with the SAP procedure.

> "*You understand, we can't, because that would mean that if some pharmacy ordered 30 different products, and only 3 were immediately available, the products would be shipped one by one when they could be; they wouldn't be grouped. And with us, you know, we have a lot of these kinds of discrepancy. So, that would mean that every day or every second day these lots would be going out. And our clients don't expect that we would work like that. And furthermore that would really be costly for us, and for the client as well. That's not the way we work at Labopharma, not at all, and it's clear that the head of commercial services and Mr. X (the CEO) would not accept that at all*" (translated freely from the original French).

As they told us, the system they already had in operation worked the way it did because it was designed to accommodate actual practice. SAP worked on a different, and, to them, incompatible, logic. But, as the interview above illustrates, it was now less clear that in the case of such discrepancies whether the company practice that would have to go, and SAP that would have priority. The down side of this latter alternative would be, in this case, much increased operating costs: a no-no from the company's point of view.

As another interview with the same sub-project head illustrates, the process was starting to look more complicated: *"For sure, some things are going to change, and others will be better. And there are others that are not going to budge. It makes for a complicated mixture, all that."* The technology was bumping up, not just against established practice, but the strategic direction of the company. And that would be less easy to dislodge. The President of the company and his top managers would be directly involved.

Since the contract that Labopharma had signed with SAP had specified a maximum of 10% adaptations, given constraints of cost, time and overall coherence, the shoe now began to pinch. Especially since, as the detailed planning and implementation proceeded, a certain number of ambiguities in the technology itself were being discovered, especially where the various modules of the system intersected with each other. Not all the procedures SAP proposed for one module (corresponding to a sector) seemed to fit very well with those in an adjoining module/sector. For example, for special orders, such as office supplies, the current practice was for each sector to handle its own orders. The project intention was to use the introduction of SAP to change this, so that orders would be directly entered into the system, which would then administer them centrally. The problem turned out to be that no one seemed to be able to identify the track the invoice would now be following: how the system would recognise *who* had issued the command and *where* to send the invoice. Even the external consultant conceded that *"Yeah, you're right, that's going to be a problem for us to fix, it'll be a real problem to identify the path the invoice takes in SAP."*

A whole set of issues was thus now emerging, of which the two described above are merely illustrative instances. One insight into the nature of the difficulties they encountered is this. As long as the company had many systems, weakly integrated, each could be adapted freely to the needs of its own sector, and thus offered a flexible tool to support local practice. By implementing a *centralised* system, the flexibility would be much more limited, if only because of the need to reconcile contrasting modes of organising, even though in other respects SAP proved to be simpler than the current technology. What the planners were encountering, in other words, was a version of the local – global tension that Suchman and Winograd had argued through in the abstract. It turns out that it is no easier to work out the contrasting pressures to integration (the SAP system) and differentiation (the existing systems) in practice, than it is in theory. As a result, the sector sub-project heads and company computer experts assigned to the various groups now proposed to the project head that a number of inter-sector meetings be set up to work through the inconsistencies. They also requested that SAP re-think its policy of limited rights of access, to emphasise sector autonomy, so that they themselves could explore in greater depth the inconsistencies they were finding. But this relaxing of constraints was inconsistent with the master plan which sought to impose its own priorities, and a fixed schedule: identify and eliminate gaps, move on to the first steps of training by developing documentation, and start the transfer of data from the old system to the new. As a result, the plan and the actual operations were now no longer matching up very well: Phase 3 was initiated, for example, even though Phase 2 had not yet been completed. The typical symptoms of SAP implementation failure were starting to appear: the spectre of cost over-runs and unplanned time delays.

A controversy resulted, pitting project sub-project heads against the overall project head, and *his* computer specialists. The overall project head: "*I insist, let's be clear about this, on the principle that your profiles and your authorisations have to be restricted at the beginning, perhaps to be extended later as the need arises, because there are so many transactions in SAP that you are going to get lost, and, worse than that, you run the risk of entering the module of someone else and, by making the wrong manipulation, destroying something he has created, or something like that. There are too many risks, you won't be able to manage them.*" To which the sub-project head for commercial operations and sales replied: "*But Philippe, I think I'm speaking for everybody here. If we don't have the authorisations, how do you expect us to do the work and how are we going to learn to use the tool, and carry out the analyses and the chosen files if we can't see what is going on in the whole sequence. It's impossible, we can't work like that, and we're not going to get anywhere.*" "*OK,*" said the project head, "*that's the end of the discussion. We'll see, I'm going to think about it. But for the moment, that's the way you're gonna work.*"

But the matter did *not* rest there. Instead, the sub-project heads decided not to wait, but instead to put their heads together and organise themselves. They began to attend each other's meetings where they tested out scenarios, traded passwords and authorisations so they could access each other's systems, to better understand the global configuration of transactions that were more or less directly related to their own functions. Finally, it was the consultants who backed down. They issued passwords for each sub-head, giving them access to all the modules. The issue was resolved in practice, even though the policy had not changed. It was merely "suspended."

As it happened, these transversal collaborations were to last throughout the remainder of the project. They were, however, not always tranquil. In fact, there were instances of spirited conflicts between sectors, in part because the changes in procedure engendered by moving to SAP also implied transfers of task responsibility between sectors. Some of the basic rules and procedures that were characteristic of the company's operations were being affected. As a result, the process was both dynamic and open-ended: adaptations that seemed to work in one meeting were identified as problematic in the next, as the inter-sector implications became evident and new adaptations seemed necessary.

Another problem cropped up: the SAP descriptions of functions such as those in the purchasing department were originally written in German. Translations into French by the consultants were not always consistent from one to another, with the result that there was residual confusion about the application of terms such as "buyer" (*acheteur*) versus "purchasing officer" (*approvisionneur*).

Then there were some strictly technical issues. In Labopharma, the manufacture of phytotherapeutic products such as jellies, pills, syrups, creams, etc. necessitates highly refined measurements of the plants and powders that compose them. The company had earlier developed specialised software that supported these measurements with an accuracy of up to 9 decimal points. SAP-RITS, however, although also developed for the pharmaceutical industry, only permitted measurements of up to 5 decimal points. Here is the reaction of one of the sub-heads to this discrepancy: *"That, perhaps you don't realise, but it is a catastrophe for us. That's going to be an enormous change, and it is going to have to be dealt with. More than that, we're going to have to find a solution, and that is going to take time. And here we are, just three months away from the official switch-over, and now we learn that it is not acceptable."*

Even the consultants were now being forced to concede that *"RITS has exploded."* The initial work plan was looking more and more unrealistic. The transition to Phase 4, in March of 2002, went by unnoticed at the working level, even though it was still the official version of what was happening, for other audiences. As one sub-head remarked: *"The transition to phase 3, 4 or 5, that's just consulting, and the management of the project for the outside, to satisfy the senior managers, and give them something to hang onto. But the reality is that it is all the phases all at the same time. No, haven't you noticed, it's a shambles"* (*laughs*).

On June 15, training activities, already underway, were suspended, and the official switch-over to the new system, foreseen to occur on August 2, was postponed to November 2002. The "big bang" was now looking suspiciously like a "whimper." There were still many problems: in August, for example, only three months away from the new official launch date there were still 543 issues in a state of suspension, as yet unresolved. Even three weeks away from the November start, no full test of the system had yet been completed because of questions of data transfer, and other technical difficulties, such as frequent server failures, as well as the persistent issue of user profiles and authorisations. What accounted for the system crashes? Which problem could be traced to issues of functionality? It was getting harder and harder to sort out the source of all the difficulties. The situation was rife for finger-pointing and assignment of blame, and indeed, we observed, it was not hard to find examples of such second-guessing.

In October, the issue was once more dumped in the lap of the steering committee. The launch was again delayed, this time to January, 2003. Gradually, however, the various contributors to the process had begun to work through the necessary compromises. In some cases, the SAP standard would prevail, with adaptations to current practice, and sometimes the solution was to find ways to get around the system, by "fooling" it in order to retain the established modes of organising. In other cases, the solution would be to construct an interface that would continue operation of the information system in place, by translation of its output into SAP, and vice versa. As a result of this compromise, some 5 of the original 13 existing technologies were actually retained.

The original previsions of the project had been down-scaled to a more realistic compromise. The software tool was itself being viewed more realistically, as well as its adequacy in meeting the needs of the company for the kind of operation it was engaged in. One consolation: one of Labopharma's main competitors, physically located nearby, had also embarked on its own ERP project. Three months after its implementation it had managed to shut down the whole production unit! Their production staff were literally thrown out of work, and the company was obliged to announce financial losses, alleviated by the hope of being back in production in three months. Labopharma personnel breathed a collective sigh of relief that they had somehow dodged a bullet.

The complexity of the system itself, they had discovered, never mind that of the organisation, precludes any easy solution to the implementation problem. Both Labopharma staff and the consultants, moreover, now had no choice but to acknowledge that the learning process they had been submitted to, as Orlikowski (1992) argued was inevitable, would not end with the official implementation. Even afterwards, it would still be a work in progress, with more adaptations still to be worked out. That implementation, in the meantime, had in any case now been delayed until March 2003. Our own role in the project, as embedded observer, ended a week later, in late March, after 18 months, 4 days a week spent in close proximity to the teams, having sat in on their meetings, and, as participant observer, become intimately familiar with their problems in the course of uncounted formal meetings and informal conversations, supplemented by continued observation and recording and familiarisation with the background documentation.

5.4 A Reconciliation of Texts?

King (1995, cited earlier) described the Suchman – Winograd debate as one more episode in "an ancient conflict over speech vs. writing" (p. 52), dating all the way back to Plato. In analysing the Labopharma experience, we want to problematise that so-called "conflict." The issue, we will argue, is *not* the tension between speech and writing, but is explained otherwise, as a confrontation between incompatible speech-and-writing, text-and-conversation configurations: between, for example, those of SAP and those of its client. It is the competing texts, and the usual conversations that they sustained and in turn sustained them, that had to be reconciled. It was not simply a speech – writing tension, even though the striking textual inconsistencies that had become evident inevitably had to be negotiated through interactive speech. In this section, therefore, we first take up for a brief examination the conversation/text relationship, to argue that conversation and text are not *different* phenomena, but are better conceived of as contrasting perspectives on the *same* phenomenon. We then focus in on one encounter of Labopharma officers and consultants to illustrate the boundary that divides incompatible text/conversation composites, each grounded in a different community of practice. The incompatibility, we will claim, is why their attempted fusion creates turbulence at the boundary between them. Finally, we suggest some of the pragmatic implications of our analysis, which may result in the eventual reconciliation of such border disputes by a progressive constitution of what we call a *meta* conversation/text.

5.4.1 Are Conversation and Text Different Modalities of Communication, or Merely Different Perspectives on it?

The confusion we have just referred to, Ricoeur (1991 [1986]) has argued, arises because we tend to confuse the text/conversation dichotomy with another, language/discourse. Language, as the classical tradition of Saussure and Hjelmslev had long before demonstrated, contrasts with speech or discourse because it is a code. It has no coordinates of time or space because it exists only as a potentiality. It is, to cite Ricoeur, "virtual and outside of time" (p. 77), "a prior condition of communication for which it provides the codes" (p. 78). Discourse, in contrast, occurs as an event: "something happens when someone speaks" (p. 77). It occurs temporally, *in* time, and in a place; it is spoken or written by *someone*, a subject; it is always *about* something (it describes, expresses, represents); it supposes the presence (immediate or virtual) of another person, an *interlocutor*. It is, Ricoeur further observes, an event in the sense that is "the temporal phenomenon of exchange, the establishment of a dialogue that can be started, continued, or interrupted" (p. 78).

But discourse must also, Ricoeur writes, be understood in another way: "if all discourse is realised as an event, all discourse is understood as meaning. What we wish to understand is not the fleeting event but rather the meaning that endures" (p. 78).

It is this notion of meaning that we need to examine closely. Here we have to be extremely careful. The meaning of a segment of discourse, one individual speaking, for example, is often taken to be self-contained (this, by the way, was one of the limitations of the speech act theory that inspired Winograd's and Flores' *The Coordinator*). We are asked, under this interpretation, to concentrate on what it, the speaking or writing, is about: what it "describes, expresses, represents." Or we focus on the subject, and his or her intentions in speaking and writing. We read motive, reason, attitude into what is said or written. Or, like much of the psychological literature on attitudes and opinion formation, we try to isolate the "effects" of a certain instance of speaking, writing or other symbolic form of representation on its hearers or readers.

What tends to get lost in these manners of representation is that the event of speech or writing is more than the *establishment* of a dialogue. It is an occurrence, among others, *in* a dialogue. But the dialogue, because it is ongoing in time, because it supposes a continuity of participants who engage in it, and because it supposes common objects of interest (*"points de repère"*), also presupposes a community and a practice that the community shares. The people within such a community do not understand each other because they speak the same *language* (although that is a *sine qua non* for the maintenance of their communication), but because they have acquired a *dialect* or specialised variant of that language that demarcates them from other speech communities (Thibault, 1997: 125–130). They do not have to look up, for example, the meaning of the word "buyer" in a dictionary (although presumably the consultants who translated SAP from the original German might have had to do so). They use the word "buyer" the way they do, and give it the meaning it has for them, because they both hear and use it daily in their discourse. They know what objects and practices it has attributed to it, and they understand the constellation of offices and configurations of authority that embed it. When they use the term "buyer" in conversation with colleagues they can be fairly confident it will be understood without ambiguity. The term and the practice it both designates and empowers are mutually constitutive.

To use a metaphor suggested by Weick (1979), the "map" and the "territory" are in reasonable alignment, as unequivocal as such correspondences can ever be. Here the "text" and the "conversation" are no more than contrasting perspectives on a single lived reality. The conversation, after all, is *itself* constituted as a sequence of textual materialisations, as Ricoeur argued. As Halliday and Hasan (1989: 10) have observed: "any instance of living language that is playing some part in a context of situation, we shall call a text." (By "living language" they mean discourse). And the conversation, in turn, must be understood as a text: we do not, in practice, laboriously decode what is said or written, syllable by syllable, or word by word, or even sentence by sentence. We grasp the patterning of discourse as a *text*, a whole that is comprehensive enough to carry meaning for us (Bruner, 1991). The issue is not whether it is spontaneous and verbal, or meticulously constructed and written. That is an important distinction in and of itself, but it is a different distinction.

What is crucial is the link between the text/conversation and its grounding in a certain practice, used by members of the community that relate to that practice. When text is used, as it is often, to bridge communities, it will not lose *all* meaning; what will be corrupted is *the meaning it had for its community of origin*. It will be assigned new meanings. Recovering the meaning it once had is now a challenge for hermeneutics. We can read the Bible, or the Torah, or the Qu'ran, but we can never transport ourselves back into the societies where they originated, nor can we can ever quite recapture the original meaning they had for the people in those different worlds of experience.

In making this argument, we do not intend to understate the importance of the distinction between speaking and writing that King, and others before him such as Ricoeur, have insisted on. It is obvious that the text that is written down supports and constitutes a very different conversation from that which unrecorded speaking leads to. One kind of conversation (the kind that Suchman had in mind) is local, situated, continuing and tightly coupled. The other is extended in time and space, links different situations, is typically sporadic and loosely coupled (Weick, 1985). Both are characteristic dimensions of the larger organisational experience of communication in organisations that grow as large as Labopharma. And, indeed, much of the turbulence that this company experienced in the course of its ERP implementation can be explained as an absence of good fit between the extended conversation, linking it to SAP with its community of designers and engineers, and the usual everyday conversations to be found in a successful enterprise.

With this in mind, let us now return to Labopharma.

5.4.2 Buying, Procuring or Purchasing? Whose Categories?

One advantage of the ethnographically inspired research we conducted into the SAP implementation in Labopharma is that it allows us, to use the image of a camera, to take a broad overview of the unfolding of the project or, alternatively, to focus in tightly on a particularly significant event. By such shifts of perspective, different facets of reality are made salient. In this section, we illustrate and comment on the kind of focusing we mean. We look at an extract of actual conversation that occurred quite early on in the project, in the autumn of 2001. It takes place in a small meeting room. At the head of the table is one of the SAP consultants. Seated at the table, on one side, is one of the eight operational heads, responsible for his sector. For purposes of identification, we will call him Gilles. Seated beside him is his second in command, Mela (short for Melanie). Across the table is the company computer specialist for the same sector, the purchasing department, Alfred (all fictional names). At the rear of the room is the researcher, Sandrine (not a pseudonym). The consultant, Paul, is facing a screen on which he is projecting a PowerPoint presentation that outlines the features of SAP the others will need to learn in order to implement the new system.

Paul: There, that's the MIGO transaction, what the purchasing agent initiates when he has completed checking out the purchasing order.

Gilles: But wait, I don't understand. He "initiates" it … what does that mean? That's already several meetings we have had about this module and the way you are talking about the purchasing agent, that's not the way it is done here, not at all. How can the purchasing agent who is supposed to look after the requested purchase, how can he initiate this transaction … That seems to suppose that it is he who takes the decision. Here, with us, I'm not sure that it works like that. What do you think, Mela?

Mela: Oh, let's see. I'm thinking about Noëlle, when she does that. No, no, it's not exactly like that. For us the problem is that it is not the purchasing agent who initiates the order, he merely enters the order into the system, he does the entry of the order. So there (turning to Paul), according to you, it's the purchasing agent who issues the order, is that right?

Gilles: There, you see, that's what I thought. You, when you speak of the purchasing agent, but it's not like that here, it doesn't have the same meaning, here with us the buyer is not the same as the purchasing agent, it doesn't have the same meaning, it's not the same function. Here, in our operation, the purchasing agent doesn't do the negotiation, it's split up into two. It's the buyer who does the negotiating. And then the buyer enters the orders in Page [an existing software] and then into Skep [another software]. Do you understand? That's why we couldn't understand the logic, we couldn't grasp it.

Alfred: I think it would have been a good idea to make a glossary of terms before we started working on this, because look it seems like there are a number of things that have the same term here, with us, and in SAP, but that don't mean the same thing. Look, we're really going to get lost this way.

Gilles: Yeah, and then the more we go on, and get into detail, the more we are going to have this problem, I think. Because it's not the first time this has happened. We need to be really clear, otherwise we're going to be wasting our time, for nothing.

Paul: Well I'm sorry, but you'd better get used to it, because that's the SAP terminology. But in the present case, let me know if you don't understand. [he turns back to his presentation, and pulls up a new slide on the screen] So now we're going to look at the organisation of the purchase.

Alfred: The organisation of the purchase, what's that, is it the purchasing department or the buyers, or both?

Paul: What do you mean by that?

Gilles: Well is it the purchaser or those who enter the orders, is it all one operation, or is it instead one or the other? Because here, with us, us using Page, it's not like that. Here, it involves several people. There's the administration of the orders, for example, the first thing to do is to consult the source file to check up on the contracts, it's the purchasing agent who does that, you see, he'll pick up the phone and call the supplier. He negotiates on the basis of the contract that is recorded in Page. It's like that, Page shows him the different contracts. And then he gets in touch with the buyer and transmits the order directly. It's always in a direct relationship, each time.

Mela: Are you sure? But it seems to me that Noëlle [user in the purchasing department] told me that she also had a role to play in the process.

Gilles: Yeah, I think you're right. Wait though. But I think she comes in at the end, for the PMS 400 [name of a sequence of purchasing transactions].

Paul: Well if you two can't even agree among yourselves, then!!! [spoken in a joking tone]

Gilles: Hey wait, I'll phone Noëlle, she'll tell us right away. Better to check directly with the source.

[He gets up, goes to the back of the room, and calls the person on the telephone. The others wait.]

Gilles: Yeah, that's it, you're right. So in the process we have to also add her and the PMS 400 of Page.

Paul: Okay, I understand.

Gilles: For us, if you like, the problem we have now with respect to purchases, is that we would like a better, clearer management of the orders. You see, there aren't many things that are automated, so we don't have all the information we need. But, on the other hand, the idea is not to automate the purchases order all the way.

Mela: That's right. You understand, purchasing, that's delicate. There's always a human decision in the background. We have to be careful.

Paul: Yeah okay, I understand. But first let me finish showing you the things I have. Afterwards, you can compare. OK, we're going on. [he shows a slide]. There, you can see in the file purchasing information, and the list of sources of purchasing. You see?

Gilles: Yeah, okay, that's a nice screen, not bad. It's user friendly. But you can enter just like that into the FPI [file purchasing information]?

Paul: Yeah, yeah, it's pretty flexible.

Gilles: Personally, I'd say it was even on the lax side. Wait a minute, does that does that mean if I understand you correctly that anybody can modify the FPI? So, if I pursue that line of reasoning a bit further, that also means that even if the FPI is not up to date, you can enter an order into it, is that right?

Paul:Yeah. You see SAP is not so rigid as all that. Often, it's very flexible. Afterwards, it's true that there, it could be dangerous, so it's up to you to do the organising.

Gilles: Myself, what really concerns me now, it's that all that, that's calling into question pretty much our whole organisation of the purchasing/buying procedure. That's how we see things, if you like. Okay so, SAP is flexible for some things, but not where we would like it to be. The problem is that it's getting at, after all, the very heart of the process. Apparently, SAP doesn't distinguish between the two. And then also the interface with the planning department is doubtful. Because, with SAP, the risk is to screw up ("shunter pas mal") the planners' work. We'll have to see if that makes sense or not. We're gonna have to ask ourselves if the way we organise things makes sense any more, or whether SAP, you see, can help us out in some way. But that's not a decision that we can make, among ourselves. We'll have to bring in the director of purchasing. Furthermore, for him, we've got to know if he favours more flexibility, but not so much that the FPI is so accessible to everyone, like that. In my opinion, we have to build in some more structure, some barriers. We'll really have to give that some serious thought. And then, after that, there is that whole business of the orders. For us, that's a real roadblock if we don't have them. Maybe make some more specific, but we're not going to budge on that.

Paul: Look, we're just at the beginning here, you shouldn't get too worked up. So we will have to consult the purchasing director, and then we'll see. So now you can make a note of the problem areas, just to keep track, and tomorrow we'll have to call in CA [the director of purchasing] or somebody he delegates and we'll talk it through.

The first thing that struck us in our interpretation of this segment of discourse, since it was so immediately evident, is that in one respect Suchman was right. The whole discussion *did* turn on the issue of categories. By the end of the exchange, it had to be clear to all present that the categories SAP used (which did not distinguish purchasing from buying) did not correspond to those that the Labopharma employees were used to. The respective functional responsibilities simply did not line up. Paul's first instinct was to dismiss the discrepancy as merely a minor problem to be resolved: *"Well I'm sorry, but you'd better get used to it, because that's the SAP terminology."* After all, the contract specifically stated that deviations from the SAP system would be kept to a strict minimum, and furthermore that, in case of doubt, it was SAP that would take precedence. He could hardly have foreseen ahead of time the stubborn (and clearly articulated) resistance he now encountered. What had seemed to be merely a minor variation in procedures had now become blown up to become a threat to the integrity of the organisation itself.

There is a different way we can look at this exchange, however, one that considerably enlarges our perspective on the encounter, beyond merely a controversy over categories. What we were now privileged to witness was a contest of texts: not, as Suchman intimated, dividing top management from its unfortunate victimised employees, but those same employees resisting the imposition of what must have seemed to them to be a foreign text: "not invented here." That SAP is a text should be evident enough: it is a composite of multiple programs and sub-programs, written in computer code, something that Paul could present using PowerPoint. That Labopharma is *itself* a text may seem less self-evident. On one level, it can be said that since it possessed its own technology (Page, Skep, among others), and its own documentary reference points (contracts, orders, bills, planning schedules, etc.), it already used texts. But these still point to documents that it constructed, and that served its purposes, and written procedures that had to be respected. That seems to fall short of claiming that Labopharma was *itself* a text.

In the next section, we develop an argument for the proposition that the organisation is in fact always a text, grounded in an ongoing conversation (Boden, 1994). The implementation of a new software-based information and communication technology (ICT) may constitute an authentic threat to its mode of existence *qua* organisation. SAP was not merely a procedural innovation. If we accept the idea that communication is not something that merely occurs *in* an organisation but is the very *basis* of organisation (Taylor and Van Every, 2000; Taylor et al., in press), then changing the modes of communication goes to the heart of the social system of the organisation itself.

5.4.3 The Organisation as Text

It would be generally agreed, we imagine, that an organisation is a rule-based set of transactions. The exchange between Gilles and Mela, for example, and their reason for consulting Noëlle, was triggered by their need to identify the "rule" they followed in their sector, the purchasing department. What is problematical in this way of thinking, however, is the very notion of a rule. In his *Philosophical Investigations*, Wittgenstein explored in considerable depth the concept of rule, to conclude that *saying* one is following a rule is not the same as actually *following* the rule (anyone who is familiar with politics will easily recognise the distinction). "How am I able to obey a rule," Wittgenstein asked rhetorically, and gave this answer to his own question: "This is simply what I do" (1958, p. 85^e). "When I obey a rule," he continued, "I do not choose. I obey the rule *blindly*" (1958, p. 85^e). But what is then happening when people say, as Gilles and Mela did, that this was the rule they followed? Wittgenstein's answer to this would simply be: "Obeying a rule is a practice. And to think one is obeying a rule is not to obey a rule. Hence it is not possible to obey a rule "privately": otherwise to *think* one was obeying a rule would be the same thing as obeying it" (1958, p. 81).

We can express much the same idea, not from Wittgenstein's philosophical point of view, but rather as set out by one of the most prestigious of contemporary management analysts, Karl Weick. "How can I know what I think," he has phrased his key perception many times over the years, "until I see that I say?" Not, please note, "until I *hear* that I say." Instead, "until I *see* that I say." More prosaically, he writes: "experience as we know it exists in the form of distinct events. But the only way we get this impression is by stepping outside the stream of experience and directing attention to it. And it is only possible to direct attention to what exists, that is, what has already passed" (Weick, 1995, p. 25).

The rule of purchasing that Gilles was insisting on conformed precisely to Weick's notion of retrospective sense making. Within Labopharma, certain patterns of conducting affairs had become the norm. When he consulted Noëlle he did not ask her what rule she *thought* she followed but rather how she would retrospectively describe what she habitually *did*. She presumably did follow a rule, in Wittgenstein's sense of conforming to a practice, but to carry out her function faithfully it is very doubtful she had to repeat to herself, *sotto voce*, "this is the rule I am following." She already *knew* how to do her job. What Gilles asked her for was what Garfinkel (1967) would have called her *account* of the rule she was following.

It is in this context that we understand the notion of an organisational text. The "text" of the organisation is the set of accounts of the practices that the members of the organisation engage in—how they account for what they actually *do*. Whether the text is materialised in speech or in writing, to return to King's distinction, is not the issue. The role of the text is to construct a universe of made-sense that enables the community of people who form the organisation to know, retrospectively, that they *constitute* an organisation because they recognise it as being rule-governed.

The patterns of communication that are typical of an organisation are, of course, themselves rule-governed; they are not merely ancillary to the functional task of issuing a purchase order. They are, as Ricoeur emphasised, activities in and of themselves. Because communication is *itself* a rule-governed activity involving people, it too can only be understood retrospectively by its subsequent translation into an account: meta-communication (Watzlawick et al., 1967). As a consequence, it is not only the way people construct their external world, *through and in their texts*, but also how they deal with it: how they negotiate a contract, for example. They also construct *themselves* as persons, with identities as members of the organisation: Gilles, Mela, Noëlle, Alfred, Paul, all with their identifiable roles and identities. As Weick has also written, "Identities are constituted out of the process of interaction. To shift among interactions is to shift among definitions of self"

(Weick, 1995, p. 20). Gilles in his conversation with Paul, for example, assumes a further role and identity, as spokesperson for his company.

The "selves" that are thus forever in the process of reconstruction, however, are not limited to the individual actors that are the ones who engage directly in communication; but they include the organisation itself. Indeed, the identity of individual members is conditional on that of the organisation, fully as much as the corollary.

Again, the dialogue illustrates this interdependence. It emerges in the use of pronouns. Consider Gilles' and Mela's first interventions, this time highlighting their use of pronouns (we have added in parentheses the original French terms, since the use of pronouns is variable from one linguistic community to another, and even within members of a group who speak the "same" language).

Gilles: But wait, **I (je)** don't understand. **He (il)** "initiates" it … what does that mean? That's already several meetings **we (nous)** have had about this module and the way **you (tu)** are talking about the purchasing agent, that's not the way it is done here (chez **nous**), not at all. How can the purchasing agent who is supposed to look after the requested purchase, how can **he (il)** initiate this transaction … That seems to suppose that it is **he (il)** who takes the decision. Here, with **us (nous),** I'm not sure that it works like that. What do **you (tu)** think, Mela?

Mela: Oh, let's see. **I'm (moi je)** thinking about Noëlle, when **she (elle)** does that. No, no, it's not exactly like that. For **us (nous)** the problem is that it is not the purchasing agent **(il)** who initiates the order, **he (il)** merely enters the order into the system, **he (il)** does the entry of the order. So there (turning to Paul), according to **you (tu)**, it's the purchasing agent **(il)** who issues the order, is that right?

Consider the structure that is implied in these comments. On the one hand, there is "us" (**nous**). Explicitly, that includes Gilles, Mela, Noëlle and an unspecified "he" (**il**). On the other hand, there is "you" (**tu**). The underlying dialogue thus links two corporate actors, "us" and "you": Labopharma and SAP. The "you" is explicitly expressed using the pronoun "tu" which, in French, is reserved for the second person singular. But as Paul's intervention slightly later indicates, he sees himself as the spokesperson for SAP. As such he is also a "nous" even though, in this incarnation, he is, in his corporate identity, a "vous" to the members of Labopharma. (The distinction between second person singular and plural has been eroded in English, but still operates in most if not all dialects of French, although with variations.) Similarly, Gilles has no hesitation in shifting identities between singular and plural: his "tu" when addressed to Mela is person to person; his "tu" addressed to Paul takes on a corporate edge, since now he speaks as a surrogate "us" (**nous**).

The dialogue has the character of a polemic. Paul informs his listeners that these are the rules of procedure they will follow: Gilles and Mela debate his interpretation, citing, as Wittgenstein might well have done himself, actual practice. Having arrived at an impasse, Gilles summons an external authority to buttress his position: first, Noëlle, because she knows what the current practice is, and then the head of the purchasing department, because it is he who will have to make the decision. Paul, for his part, calls (although discreetly) on a different authority, that of SAP.

The issue is clear: which text takes precedence. The authority resides in that office that can choose the right text, and thus effectively "author" it.

5.4.4 The Role of Conversation

We cited Halliday and Hasan's definition of text as a string of language that is "doing some job in some context." The context they had in mind is communication. Put somewhat differently, a text is not a *text* because someone spoke or wrote it. Nor is it a *text* because someone interpreted it: heard or read it. It is a text because it is both authored and read. Only in this way can it do its "job." What we were observing in the long ongoing conversation linking Labopharma and its system supplier over a 2-year period was, to start with, a confrontation of texts, and the tissue of identities they mediated, that progressively transformed itself into a new text, one that now reflected a modified of practices. It took time, but eventually compromises were worked out, and the situation stabilised. It was not that, first, they negotiated a new text, and then applied it to practice. Nor was it that the practice changed, but the text remained. Text and the practice of conversation are not autonomous phenomena. They are mutually generative. As Giddens (1984) has observed: "Human action occurs as a *durée*, a continuous flow of conduct, as does cognition ... "Action" is not a combination of "acts": "acts" are constituted only by a discursive moment of attention to the *durée* of lived-through experience" (Giddens, 1984, p. 3). The only way people can make sense of their own conversations, he is intimating, is by "texting' them. Vice versa they are not even texts until they constitute the matter of a conversation, following Halliday and Hasan.

That prolonged conversation was, however, itself more than merely working through the intricacies of adapting the technology to the work habits of the company, and vice versa. That is the pointing-outward dimension of communication. It was simultaneously inward-pointing: a tacit contest to establish, under threat, who had authority, and who did not. We have already observed how Gilles progressively backed Paul into a corner, and, in doing so, claimed primacy for the authority of his own group, and its officers. What will be less immediately evident, in a rapid reading of the transcript, was a secondary instance of by-play, this time involving Paul and Alfred. Paul, in his role, was relatively junior, without as yet a long accumulated experience as a consultant. Alfred, although identified as a computer expert in the company, had himself previously followed a career as a consultant, and enjoyed considerable respect for his accumulated know-how. When he quietly remarked that Paul might have thought to assemble a glossary of terms before his briefing he was assuming an identity as someone in Paul's own area of expertise—someone with superior qualifications. The only weapon Paul possessed was to protest that *"you'd better get used to it, because that's the SAP terminology."* Beyond that, his personal authority was insufficient to carry the day.

An organisation is a system of authority. A prime purpose of all conversation is to maintain and establish the authority pecking order. If there is an innovation, neither calling up the official text, nor the "old" text, is any longer effective for exactly the reason that Wittgenstein identified: it is no longer the *right* text.

Considered by some authors as the reason for many implementation problems, and by others as the condition for a good adoption, the adaptation of an ERP to specific needs is a key issue of the integration of the system in the organisation. We shall first try to be more precise on the various levels of adaptation which are possible, then discuss what can really be expected from adaptation.

5.5 Conclusion

The paradox of the computerisation of modern enterprise is that, over the some 60 years since the modern computer became operational, the character of work, and the practices of administration, have been totally transformed by its introduction. And yet, on a project by project basis, the organisational landscape is littered with the detritus of as many failures as successes. At the beginning of this chapter, we cited two contrasting views on the experience of computerisation, as organisational members experience it. Suchman saw it in stark terms, as the imposition by one privileged group of its "categories" on a hapless working population. That is not what we observed at Labopharma. There, it was the dedicated employees who finally worked through the necessary modifications of the proposed system, not out of slavish obedience, but because they themselves understood, as Winograd had argued, the importance of a reliable tool for handling such operations as purchasing. On the other hand, Winograd saw the goal of computerisation as a standardisation of communicative practices, right down to the level of the ordinary conversation. The people we were observing and talking to had no intention of underplaying the role of individual judgment and discretion, even though they recognised the need for a clearer, more systematic definition of some of their processes of work.

Our own reading of the difficulties so many projects encounter is that designers and marketers of systems such as ERP regularly tend to overlook the simple fact that an organisation, in its very essence, is not merely an assemblage of practices, to be redesigned. It is a text—and a text that carries authority for those who work there. The developers still think of communication as what goes on in an organisation. What they fail to understand is that the organisation is itself a communicative construction. If the system is to work it is not merely the practices that must change. The text must also be rewritten. But the text is not merely an accounting of practices. It is also a tissue of identities: lived, not merely transcribed to text. To transform people's identities, arbitrarily, is a much more delicate matter than simply altering a few practices. The system of authority that is legitimated by its text, and is the backbone of the organisation, is also threatened.

Labopharma was fortunate in having survived the transition, relatively unscathed. Its closest competitor, geographically, was not so fortunate.

5.6 References

Barley S, (1996) Technicians in the workplace: Ethnographic evidence for bringing work into organisation studies. Administrative Science Quarterly 41(3):404–440

Beatty RC, Williams CD, (2006) ERP II: Best practices for successfully implementing an ERP upgrade. Communications of the ACM 49 (3):105–109

Boden D, (1994) The business of talk. Cambridge, UK: Polity Press

Brown JS, Duguid P, (2000) The social life of information. Boston, MA: Harvard Business School Press

Bruner J, (1991) The narrative construction of reality. Critical Inquiry Autumn:1–21

De Michelis G, (1995) Categories, debates and religion wars. Computer supported cooperative work (CSCW) 3:69–72

Eisenberg E, (2007) Strategic ambiguity. Thousand Oaks, CA: Sage

Garfinkel H, (1967) Studies in ethnomethodology. Englewood Cliffs, NJ: Prentice-Hall

Giddens A, (1984) The constitution of society. Berkeley and Los Angeles: University of California Press

Goffman E, (1959) The presentation of self in ordinary life. Garden City, NY: Doubleday

Grudin J, Grinter RE, (1995) Ethnography and design. Computer supported cooperative work (CSCW) 3:55–59

Halliday MAK, Hasan R, (1989) Language, context, and text: Aspects of language in a social-semiotic perspective. Oxford, UK: Oxford University Press

Hoppenbrouwers S, (2003) Freezing language: Conceptualisation processes across ICT-supported organisations. PhD thesis, Catholic University of Nijmegen, the Netherlands

King JL, (1995) SimLanguage. Computer supported cooperative work (CSCW) 3:51–54

Lawrence PR, Lorsch JW, (1969) Organisation and environment: Managing differentiation and integration. Homewood, IL: Irwin

Malone TW, (1995) Commentary on Suchman article and Winograd response. Computer supported cooperative work (CSCW) 3:37–38

Morgan G, (2006) Images of organisation (rev. version of 1986 book). Thousand Oaks, CA: Sage

Orlikowski W, (1992) The duality of technology: Rethinking the concept of technology in organisations. Organisation Science 3:398–427

Ricoeur P, (1991) From text to action: Essays in hermeneutics, II (trad., K. Blamey & J. B. Thompson). Evanston, IL: Northwestern University. (Originally published as Du texte à l'action: Essais d'hermeneutique, Paris, Seuil, 1986)

Shannon CE, Weaver W, (1949) The mathematical theory of communication. Urbana, IL: University of Illinois Press

Suchman L, (1994) Do categories have politics? The language/action perspective reconsidered. Computer supported cooperative work (CSCW) 2:177–190

Suchman L, (1995) Speech acts and voices: Response to Winograd et al. Computer supported cooperative work (CSCW)3:85–95

Taylor JR, Casali A, Marroquin L, Vasquez C, (in press) O essencial sobre comunicacão organisacional. Coimbra, Portugal. Angelus Novus

Taylor, J. R., Cooren, F., Giroux, N. & Robichaud, D. (1996). The communicational basis of organisation: Between the conversation and the text. Communication Theory 6 (1):1–39

Taylor J.R,Van Every E, (2000) The emergent organisation: Communication as its site and surface. Mahwah, NJ: Lawrence Erlbaum Associates

Thibault PJ, (1997) Re-reading Saussure. London and New York: Routledge

Watzlawick P, Beavin JH, Jackson DD, (1967) Pragmatics of human communication. New York: W. W. Norton

Weick KE, (1995) Sensemaking in organisations. Thousand Oaks, CA: Sage

Weick KE, (1985) Sources of order in underorganised systems: Themes in recent organisational theory. In Y. S. Lincoln, ed., Organisational theory and inquiry:106–136. Beverly Hills: Sage

Weick KE, (1979) The social psychology of organising. New York: Random House

Winograd T, (1994) Categories, disciplines, and social coordination. Computer supported cooperative work (CSCW) 2:191–197

Winograd T, Flores F, (1986) Understanding computers and cognition: A new foundation for design. Norwood, NJ: Ablex

Wittgenstein L, (1958 [1953]) Philosophical Investigations. New York: Macmillan

Wittgenstein L, (1974) Tractatus Logico-Philosophicus. New York and London: Routledge & Kegan Paul (tr. D.F. Pears & B.F. McGuinness). Originally published in 1921 in Annalen der Naturphilosophie

6

Contradictions and the Appropriation of ERP Packages

Ben Light, Anastasia Papazafeiropoulou
Information Systems, Organisation and Society Research Centre,
University of Salford

6.1 Introduction

Enterprise Resource Planning (ERP) software typically takes the form of a package that is licensed for use to those in a client organisation and is sold as being able to automate a wide range of processes within organisations. ERP packages have become an important feature of information and communications technology (ICT) infrastructures in organisations. However, a number of highly publicised failures have been associated with the ERP packages too. For example: Hershey, Aero Group and Snap-On have blamed the implementation of ERP packages for negative impacts upon earnings (Scott and Vessey, 2000); Cadbury Schweppes implemented plans to fulfil 250 orders where normally they would fulfil 1000 due to the increased complexity and the need to re-train staff post-implementation (August, 1999) and FoxMeyer drug company's implementation of an ERP package has been argued to have lead to bankruptcy proceedings resulting in litigation against SAP, the software vendor in question (Bicknell, 1998). Some have even rejected a single vendor approach outright (Light et al., 2001). ERP packages appear to work for some and not for others, they contain contradictions. Indeed, if we start from the position that technologies do not provide their own explanation, then we have to consider the direction of a technological trajectory and why it moves in one way rather than another (Bijker and Law, 1994). In other words, ERP appropriation cannot be pre-determined as a success, despite the persuasive attempts of vendors via their websites and other marketing channels. Moreover, just because ERP exists, we cannot presume that all will appropriate it in the same fashion, if at all. There is more to the diffusion of innovations than stages of adoption and a simple demarcation between adoption and rejection. The processes that are enacted in appropriation need to be conceptualised as a site of struggle, political and imbued with power (Hislop et al., 2000; Howcroft and Light, 2006). ERP appropriation and rejection can therefore be seen as a paradoxical phenomenon. In this paper we examine these contradictions as a way to shed light on the presence and role of inconsistencies in ERP appropriation and rejection. We argue that much of the reasoning associated with ERP adoption is pro-innovation biased and that deterministic models of the diffusion of innovations such as Rogers (2003), do not adequately take account of contradictions in the process. Our argument is that a better theoretical understanding of these contradictions is necessary to underpin research and practice in this area.

In the next section, we introduce our view of appropriation. Following this is an outline of the idea of contradiction, and the strategies employed to "cope" with this. Then, we introduce a number of reasons for ERP adoption and identify their inherent contradictions using these perspectives. From this discussion, we draw a framework, which illustrates how the interpretive flexibility of reasons to adopt ERP packages leads to contradictions which fuel the enactment of appropriation and rejection.

6.2 Views of Information and Communications Technology Appropriation Processes

Innovation is an idea, method or device that is perceived as new by those in the social system where it is manifested. Innovations may be products, such as fax machines, techniques such as structured programming practices or even social reforms (King et al. 1994). Therefore, ERP packages can be viewed as innovations because they can be perceived as a new product (SAP), service (their functionality) or social reform (their so called inscribed "process orientation"). Although Rogers' Diffusion of Innovations (DOI) theory (Rogers, 1995; Rogers, 2003) is one of the best known thesis in the area of adoption and has been used and extended widely, it has also been heavily criticised (Kautz and Pries-Heje, 1996; Allen, 2000; Elliot and Loebbecke, 2000; Lyytinen and Damsgaard, 2001; Papazafeiropoulou, 2002). Rogers aims to trace and explain the path of an innovation's acceptance through a given social system, over time and according to his thesis, there are perceived attributes of innovations which affect this. Rogers pays a lot of attention to these attributes in his work and indeed, other studies have sought to extend these further (Moore and Benbasat, 1991; Agarwal and Prasad, 1997). However, although Rogers refers to social influences that may impede or facilitate the process, the emphasis tends to be on the innovation itself. The hyphen in the socio-technical remains in that the "technology of innovation" is bracketed off from influences – such as what he calls the promotional efforts of change agents. Clearly this is problematic if one takes the position that the technical and social are negotiable in nature (Bloomfield and Verdubakis, 1994). We agree with Beynon Davis and Williams (2003) and O'Neill et al. (1998), who criticise the rational account of technological diffusion, particularly as they argue that environments and the role of various actors during the appropriation process need to be further emphasised. We believe that complex networks of actors and their conflicting ideas or requirements can influence the appropriation or rejection of an ICT in unpredictable ways.

A social shaping perspective emphasises technological development as a socio-technical process. It critiques and transcends social and technologically deterministic accounts of appropriation (Sørensen, 2002). Thus in terms of ICT appropriation, we can draw on social shaping to suggest that it is less than certain the way it will play out. Social shaping approaches emphasise the way that technologies are configured throughout the appropriation by various people in different social groups (Bijker and Law, 1994, Fleck, 1994). Technological development is not a linear process with one possible outcome, rather a process during which the form of an artefact becomes "stabilised" as consensus emerges among key social groups. Thus, such accounts are not restricted to the social groups of design-room engineers or laboratory personnel (Bijker, 1994). Thus we attend to the idea of "Relevant social groups", those who share a meaning of an artefact and, including for example, engineers, advertisers and consumers (Kline and Pinch, 1999). For this study, a number of relevant social groups have been identified by "following the actors" and "historical snowballing" in line with (Bijker, 1994). These are depicted in Figure 6.1. As Bijker suggests "this is of course an ideal sketch as the researcher will have intuitive ideas about what set of relevant social groups is adequate for the analysis of a specific artefact" (Bijker, 1994: 77). Individuals can be members of more that one group and we do not take any group to be homogenous. Clearly, there will be great differences among those in our groupings but they are useful for aiding the forthcoming analysis. The identification of relevant social groups, links to the ideas of interpretive flexibility and closure (Russell and Williams, 2002). Interpretive flexibility refers to the scope for the attribution of different meanings to an artefact, according to the different backgrounds, agendas, purposes and commitments of those groups and/or individuals. Closure refers to the process by which, or the point at which, interpretations of an artefact are brought into agreement, or whereby one interpretation becomes dominant. Thus, social shaping can be helpful in surfacing and explaining contradictions. In the next section we briefly outline the idea of contradiction and the strategies employed to cope with this.

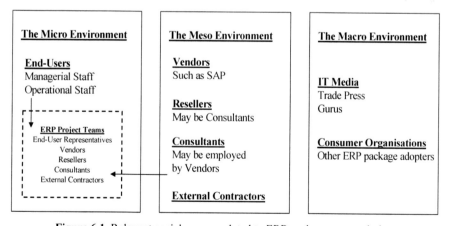

Figure 6.1. Relevant social groups related to ERP package appropriation

6.3 Coping with Contradictions and ERP Packages

In the introduction we indicated that ERP packages are inherently contradictory. By this, we mean that ERP packages embody tensions. For example, one person might perceive a package as a more cost effective route to systems development, over custom development, where another does not. Indeed, many of the contradictions we raise naturally refer back to perceptions of custom development. It is important to understand such processes of contradiction making and deployment as they are, we would argue, a central feature of appropriation. For example one could consider the work of Boudreau and Robey (1996) regarding the contradictory nature of BPR. They argue that theories which incorporate a logic of contradiction are valuable for studying such things as organisational change, an activity often linked with technology appropriation. They state "In such theories, opposing forces interact to cause resulting solutions that are only partially predictable". The dynamic and nondeterministic nature of such theories may frustrate conventional attempts to produce and validate causal models, but such theories have definite advantages for explaining complex phenomena such as organisational change" (Bourdreau and Robey, 1996: 54). The study of contradictions is also interesting as such instances can be viewed as a potential basis for insights into conflict within and between social groups (Walsham, 2002).

What is also of interest is how people cope with contradictions. Occupational life is one such area that has been studied as inherently contradictory. For example, Kase and Trauth (2003) suggest three ways that women cope in the IT workplace. They say that a woman may cope by "Assimilation" whereby they deny that discrimination against women occurs. "Accommodation" involves the recognition and acceptance of discrimination as part of every day life and "Activism"', the questioning of inconsistencies and contradictions in a male dominated workplace. A different application has been to the area of BPR. Jones (1995) suggests three strategies for coping with contradictions in BPR, Denial, Resolution and Accommodation. Those in Denial refuse to acknowledge contradictions – they dismiss these as misconceptions or the product of flawed research. Those adopting a strategy of Resolution believe that apparent contradictions are compatible and that contradictions may arise, for example, as a result of context diversity. Those who Accommodate, suggest that contradictions in BPR should be accepted as a normal feature of organisational life.

As with earlier Manufacturing Resource Planning (MRP) packages (Swan et al., 2000), ERP packages have been black boxed whereby they appear closed and not open to substantive modification. This raises difficulties as, despite the rhetoric of proponents, they cannot be inserted anywhere and this can lead to contradictions. In the following three sections we consider the contradictory nature of ERP appropriation. The first section, *the idealisation of ERP packages* is concerned with the idea that they represent an economical solution to functionality problems of existing or projected systems. The second section, *the myth of the perfect ERP service*, refers to perceptions that they "deal with" legacy information system problems because they are better built, allow implementers to adopt "best practices" and are well supported. The third category of *contradictions and relevant social groups* focuses upon the diverse agendas and influences of those involved in the processes of diffusion and adoption. Moreover, we will later go on to argue that this contradictory reasoning leads to the shaping and reshaping of reasons for adoption and rejection of ERP packages and competing technologies. Someone might of course categorise these reasons differently. Needless to say, our point here is that a group of reasons exist and that however they are formulated they will influence the process.

6.3.1 The Idealisation of ERP Packages

One argument put forward by proponents of ERP package adoption is the conception that potential adopters can now find the "right" package for their organisation. For example, many ERP vendors offer products that are "industry specific". Yet, it has long been suggested that software packages seldom, if ever, match end-user requirements exactly (Gross and Ginzberg 1984; Weing 1984). Although these studies were looking at packaged software in general, and some time ago, extracts from the Gross and Ginzberg study such as "Available packages do not adequately reflect my industry" and "My needs are too unique to be adequately represented in available packages" still have resonance today in an ERP context. For example, in one ERP study, the IT Director stated that although he felt that enterprise systems were good, his company would have to build around them (Light et al., 2001). Indeed, highly integrated sets of packages (such as ERP) may vary considerably in quality and functionality on a module-by-module basis (Andersson and Nilsson, 1996). For example at "Global Entertainment" single vendor based packages were evaluated. However, they were perceived as being historically built from packages aimed at specific functions and then expanded for enterprise coverage (Light et al., 2001). Another attribute of ERP packages that is promoted as advantageous over custom approaches is their low unit cost (PriceWaterhouse, 1996; Klepper and Hartog, 1992; Chau, 1995; KPMG, 1998). Moreover, the costs of acquisition, implementation and usage of packages are argued to be reliably predictable and lower than for custom developed software because they are posited as complete technologies (Golland, 1978; Heikkila et al., 1991). Nevertheless to implement an ERP package is not just about the price of a licence. Although the initial implementation of the software might be cheaper, further costs arise when companies start customising the packages to meet their specific company needs as at "Cable" and "Home", see (Light, 2001). As Light and Wagner (2006) and Wagner and Newell (2006) have argued, ERP packages should be conceptualised as requiring ongoing work in situ. Therefore, ERP projects might display "cost over-run", problems normally associated with custom development (Remenyi et al., 1997). At "Threads" for example, the overall project was reported to have increased in cost five-fold from original estimates (Holland and Light, 1999) and it is doubtful that FoxMeyer anticipated the ultimate costs of the acquisition, implementation and usage of SAP, which was argued to have led to bankruptcy proceedings (Bicknell, 1998).

6.3.2 The Myth of the Perfect ERP Service

Another aspect of ERP packages that is part of the pro-adoption argument is the quality of the service offered by vendors and the software. For example diffusers and adopters praise ERP packages, as having the ability to help those in organisations overcome legacy IS problems. They are argued to be: well structured and allow for maintenance and future development to be outsourced to a vendor (Butler, 1999; Scheer and Habermann, 2000; KPMG, 1998); easily operated, supported and maintained due to the ability of the implementing organisation to tap into available a skills base for the software (Bingi et al., 1999; Sumner, 2000; Willcocks and Sykes, 2000); and well documented and organised (Golland, 1978; Butler 1999; Scheer and Habermann, 2000). For example, Novartis management gave the proliferation of ad hoc systems, minimal attention to maintenance, and the lack of interoperability as the reasons for the move to ERP packages (Bhattacherjee, 2000). Nevertheless, we believe that to treat ERP packages as different to legacy information systems is inherently flawed (Light, 2003). One study highlights the irony of the belief in ERP packages as the "replacement" for legacy information systems – 41 per cent of adopters stated they were locked-in to the packages they had bought to replace "legacy" custom built programs (PriceWaterhouse, 1996). The implication of this is that although ERP packages may have diffused rapidly because of their perceived ability to relieve legacy information system problems, they may also introduce new ones. Similarly, ERP packages have been advertised as an easy way to face application backlogs due to rising software development costs and the need for rapid deployment of new systems to keep pace with strategies (PriceWaterhouse, 1996; Li, 1999). Indeed, it is further argued that the lengthy lag between a user's request for a new system and implementation (a supposed feature of custom development) has been replaced by market-based approaches where software vendors can produce new releases faster than consumers can absorb them (Sawyer, 2001). However, end-users still have to wait for the product to be built and implemented (Butler, 1999), and when they have implemented it, they may have to wait for upgrades and maintenance activities to be performed (Gross and Ginzberg, 1984; Adam and Light, 2004). For example, those at Dell decided that the deployment cycle for the SAP package would have taken them too long. Their plan, to convert all of the company's information systems to the SAP package, was estimated to require several years to implement and thus the project was abandoned (Fan et al., 2000).

For a long time, packages have been promoted as "tried and tested products", and in most cases, as having been installed by other organisations (Heikkila et al., 1991; Golland, 1978). ERP packages have been no exception, vendor websites usually contain the lists of high profile company cases that promote the benefits resulting from the implementation of their product. Yet again, there are problems with these assertions. There is the suggestion that ERP packages are "better built" than custom developed software yet it has been suggested there is a lack of rigour in the product development processes of the packaged software industry (Carmel, 1993; Carmel, 1997). In addition ERP packages are promoted as innovations that can give those in organisations access to a broader knowledge and skills base. The adoption of ERP packages for this reason is evident at the Crosfield, DMC Prints and Nokia organisations where only a few employees were capable of handling the administration and development of their existing custom developed software (Dolmetsch et al., 1998). Furthermore, the benefits of increased familiarity among the user population can also be realised. These may include opportunities for increased intra-organisational and inter-organisational knowledge sharing to enable the speedier, and more successful, deployment of packages (Pan et al., 2001; Newell et al., 2002). Problems may however, arise if a particular package becomes very popular and this may lead to difficulties for end-users in a consumer organisation being able to obtain the skills they need. Therefore, although ERP packages may be chosen to "buy into" a knowledge and skills base, difficulties may arise with "successfully promoted" and widely adopted innovations. It also follows from this that problems may also emerge if a product is, or becomes, less popular which might mean that the support for the package may be hard to find. For example, the reported shortage of Assembly skills in 1994 (Bennett, 1994) echoes the widespread lack of SAP consultants in the late 1990s – early 2000s.

Finally, the adoption of ERP packages is frequently related to the attainment of "best practices". The central theme is that there are advantages to be obtained by adopting ERP packages over similar custom development because of the ability to "buy into" the best practices that are written into the software (Klaus et al., 2000). However, questions have to be raised about the possibility of attainment of the perceived advantages to be gained from the adoption of standard best practices. As with Manufacturing Resource Planning packages (Swan et al., 2000), the forerunner to ERP packages, what may be good for one adopter may not be for another (Wagner et al., 2006).

6.3.3 Contradictions and Relevant Social Groups

Contradictions in adoption reasoning become very clear when we consider the diverse agendas and levels of influence those in various social groups invoved in the diffusion and adoption of ERP packages have (Adam and Light, 2004; Scott and Kaindl, 2000; Pozzebon, 2001). Managers in user organisations may choose to implement ERP packages with the explicit desire to force change, or use the ERP packages as the excuse for change (Champy, 1997). Yet, this argument needs to be considered carefully when used for pronouncing the change agent capabilities of ERP packages. Whilst it is clear that ERP packages have unintended consequences (Hanseth and Braa, 1998), like other innovations the extent of the change and the labelling of beneficiaries are debatable. Suchman and Bishop (2000) argue, for instance, that innovations may be used to reinforce managerial control systems rather than improve everyday working life for operational staff. As King et al. (1994) suggest, innovation is political, and its desirability varies greatly when the question of who does and does not benefit is raised. Thus ERP packages could be seen as a way to reinforce the status quo rather than do anything dramatic, depending upon your interpretation of the situation.

Another reason for the adoption of ERP packages is bravado. For example, a reason for the adoption of an ERP package in one study was "To be able to show the big boys" (Adam and O'Doherty, 2000) and in another, it was because many other chemical companies were implementing it (Ross, 1999). As one survey highlights 66 per cent of respondents agreed that "without this package we would be at a competitive disadvantage in our industry" and 50 per cent were motivated to adopt because "we were one of the first in the industry to adopt this package" (Swanson, 2003: 65–66). Thus, processes and reasons for adoption are often fuelled by bullishness and ideas of being fashionable (Kieser, 2003). This contradicts dominant existing theories of packaged software selection, which take a rational choice view whereby decision making is a function of economic impact alone (Lynch, 1984; Chau, 1994; Nelson et al., 1996; KPMG, 1998). We have to remember that internal and external Salesmanship efforts are integral to appropriation processes (Friedman and Cornford, 1989; Howcroft and Light, 2006). Those in organisations may adopt ERP packages due to an approach by a vendor, or reference to vendor publications, market surveys, the Internet, and other adopters. For example "strong ERP vendor marketing" and "The right solution and message at the right time" have been cited as key reasons for its adoption (Klaus et al., 2000). Therefore, those in organisations may be "sold" the idea of ERP packages and a particular product. Moreover, this selling activity may be linked with the exercise of, in particular, symbolic power – where an ERP package is chosen where it was not necessarily the best choice (Markus and Bjørn-Andersen, 1987; Howcroft and Light, 2006). This final category can be seen as the fuel of the diffusion and adoption process in that this frames appropriation by exerting influence upon those selecting coping strategies.

6.4 Discussion

This review of the contradictory nature of ERP package appropriation is not intended to be exhaustive. Our aim here has been to stall the closure of our understanding of ERP packages as espoused by those in a wide variety of social groups. This is particularly important as, over time, views of ERP packages will change. As has been pointed out with ICTs in general, once institutionalised, they may become taken for granted, when they break down they emerge again from the background (Silva and Backhouse, 2003). Our argument is therefore, that we need a fuller understanding of the contradictions of ERP packages as this has inextricable links with the difficulties that the various relevant social groups in the micro, meso and macro environments might experience in living with them. Rogers' DOI theory is normative in that it prescribes how the diffusion of innovations will take place. The problem with this is that there are always going to be contradictions that cannot be treated as exceptions to the rule. Although Rogers calls for studies that do not contain a pro-innovation bias, his work clearly is. His framework assumes adoption, when it could so easily be taken to be reasons against adoption. Moreover, although Rogers talks about heterogeneity in the adopter population, he assumes a homogeneous adopter and diffuser population thereby denying the possibility of contradictions.

Table 6.1. Examples of approaches to coping with ERP contradictions

Approach	Example
Denial - The belief that there are no contradictions in ERP packages.	In the case of HealthFirst, there is evidence of an implementers buying into vendor rhetoric. According to Rodgers, the Vice President and Chief Information Officer "Peoplesoft was chosen primarily because of its healthcare expertise "They understand the unique needs of our industry and they build that knowledge into their products" (Shang and Seddon, 2003: 84)
Accommodation - A recognition that contradictions exist, which are not helpful, but these are viewed as part of the trade off when adopting packaged software.	At Threads, when discussing appropriating a European wide ERP package, the Project Manager said that: "It's like building a house, you have to get the foundations right… this wasn't the time to start worrying about the carpets and curtains." and it was reported that "the objective was to achieve 90% global processes and 10% national specific ones to deal with national variances such as financial reporting, tax and customer preferences." (Holland and Light, 1999: sic)
Resolution - The recognition that contradictions exist, which are not helpful and thus they need to be challenged.	At Big Civic – the researchers reported that "the Supplier was finding it increasingly difficult to continue to resource the production of generalisable concepts, particularly when many appeared to work across a few sites only. Therefore, the universities were drawn into a struggle with the Supplier and with one another over the inclusion of their specific needs. This as described by the Project Director at Big Civic led to a "push and shove" between them and Large Campus." (Pollock et al., 2003: 327)
Acceptance - A recognition that contradictions exist, but these are accepted.	In the case of Shop floor control at Cable it was reported that although ERP packages cannot do everything out of the box. "The ERP software's progress reporting screens were too cluttered and complex. A trial of the new shop floor procedures was undertaken and the results were disappointing—17% of data entry was inaccurate. The company used the ERP software Applications Programming Interface (API) tool to simplify production progress reporting screens. Trials of this new screen reduced the error rate to 8% immediately and virtually eliminated it within two weeks." (Light, 2001: 420)
Rejection - An ERP package is not adopted.	Those at Dell decided that the deployment cycle for the SAP package would have taken them too long. The plan, to migrate all of the company's systems to the SAP package, was estimated to require several years to implement and thus the project was abandoned (Fan et al., 2000).

Rogers' work also separates the social from the technical. He treats these as variables that determine the rate of the adoption of innovations. We argue that drawing upon social shaping, a view which incorporates a logic of contradiction, blurs the boundary between these. Moreover, the concepts of relevant social group, interpretive flexibility and closure are helpful in furthering our understanding of how people cope with contradictions and how such contradictions, and the coping strategies shape processes of appropriation. To more concretely illustrate this we will now further analyse this interplay. Table 6.1 details indicative published case based examples of the contradictions in ERP adoption reasoning and the resulting coping strategies employed. Based on an analysis of the discussion so far, and the illustrative examples in Table 6.1, it is possible to discern five approaches to coping. We recognise that various coping strategies might co-exist as a consequence of various viewpoints and that an individual might adopt a different coping strategy at different times.

Thus far then, we have identified that ERP adoption reasoning is contradictory and that the flexible interpretation of their representation leads to deployment of strategies for coping with contradiction. Based upon this analysis, we have produced a conceptual framework which postulates that reasons for appropriation and rejection are socially constructed and subject to interpretive flexibility, and thus are inherently contradictory (Figure 6.2). The framework should be seen as a way for negotiating a shared understanding of such processes, not the way they are pre-determined to play out.

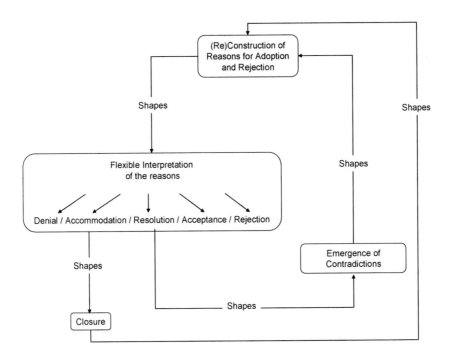

Figure 6.2. A framework for analysing contradictions and ERP appropriation

To explain the framework, we suggest that an initial set of grounds for the appropriation and/or rejection of ERP packages are constructed. In this paper, we have categorised these reasons into three groups as discussed earlier. These reasons are then subject to representation and interpretation which takes the form of the enactment of various coping strategies. This will then lead to the emergence of contradictions. These contradictions then inform the further construction of, potentially, a revised set of reasons that are, once again, subject to representation and interpretation. Thus, the coping strategies shape and are shaped by the contradictions. Another important feature of the framework is its implicitly longitudinal view of appropriation. Within this process, there is the opportunity for some form of closure to be attained whereby certain groups share a view of the technology in question. This of course may also be part of the mutually constitutive process of responding to the construction of contradictions. The framework leaves space for the reopening of the technology based on such activities as the reconstruction of existing reasons for adoption and rejection and the creation of new ones. Such new reasons may be based on the shaping efforts of social groups targeted at the reasons themselves or their efforts in developing alternative innovations – ERP II for example. Also, although we discuss adoption in isolation here, this process is an inherent part of the processes of selection, implementation and usage of ERP packages. Those in the micro, meso and macro environments can build and "knock down" contradictions, and hitherto reasons for adoption and rejection at any point in time.

6.5 Conclusion

ERP packages have been widely adopted in recent years but there is limited work that focuses upon the contradictions related to their adoption by organisations. We have attempted to tackle the pro-innovation bias present in many of the reasons reported for the adoption of ERP packages and deconstructed these using a social shaping view. As with custom development, we show that the reasons for ERP package adoption cannot be viewed as singularly rational. Some of the reasons for ERP package adoption have mythical qualities, or at the very least, are questionable. We use social shaping to demonstrate the importance of interpretive flexibility in understanding how relevant social groups might cope with ERP contradictions and how these interpretations can lead to a "closure" of the appropriation process. Our work points to the roles that social groups play in constructing and shaping the reasons why those in organisations should (or should not) adopt ERP packages over time. We propose a framework where coping strategies fuel the appropriation process over time and believe that its application can help in better understanding of the complexity of ICT appropriation more generally. In particular many of the reasons could be considered in relation to other packaged software applications and there are elements of our work that even have resonance for custom development especially as this can also have a market-oriented dimension to it.

6.6 References

Adam F, O'Doherty P, (2000) Lessons from Enterprise Resource Planning Implementations in Ireland – Towards Smaller and Shorter ERP Projects. Journal of Information Technology 14(4):305–316

Adam A, Light B, (2004) Selling Packaged Software: An Ethical Analysis. In Proceedings of the 12th European Conference on Information Systems Turku, Finland

Agarwal R. Prasad J, (1997) The Role of Innovation Characteristics and Perceived Voluntariness in the Adoption of Information Technologies. Decision Sciences 28(3): 557–582

Allen JP, (2000) Information Systems as Technological Innovation. Information Technology and People 13(3):210–221

Andersson R, Nilsson AG, (1996) The Standard Application Package Market – An Industry in Transition? Advancing your Business: People and Information Systems in Concert. M. Lundeberg and B. Sundgren. Stockholm, EFI, Stockholm School of Economics:1–24

August V, (1999) ERP Sites Hit by Performance Dip. Information Week 17, February:12

Bennett K, (1994) Legacy Systems: Coping with Success. IEEE Software 12(1):19–23

Beynon Davis P, Williams MD, (2003) The Diffusion of information Systems Development Methods. Journal of Strategic Information Systems 12(1):29–46

Bhattacherjee A, (2000) Beginning SAP R/3 Implementation at Geneva Pharmaceuticals. Communications of the AIS 4(2):1–39

Bicknell D, (1998) SAP to Fight Drug Firm's $500M. Suit Over R/3 Collapse. Computer Weekly 3 September:3

Bijker WE, Law J, (1994) Shaping Technology/Building Society: Studies in Sociotechnical Change, MIT Press, Cambridge, MA

Bijker WE, (1994) The Social Construction of Fluorescent Lighting, or How and Artefact was Invented in Its Diffusion Stage. In W. E. Bijker and J. Law., Shaping Technology/Building Society: Studies in Sociotechnical Change. MIT Press, Cambridge, MA:75–104

Bingi P, Sharma MK, Godla JK, (1999) Critical Issues Affecting an ERP Implementation. Information Systems Management 16(3):7–14

Bloomfield BP, Vurdubakis T, (1994) Boundary Disputes: Negotiating the Boundary Between the Technical and the Social in the Development of IT Systems. Information Technology & People 7(1):9–24

Boudreau M, Robey D, (1996) Coping with Contradictions in Business Process Reengineering. Information Technology and People 9(4):40–57

Butler J, (1999) Risk Management Skills Needed in a Packaged Software Environment. Information Systems Management, 16(3):15–20

Carmel E, (1993) How Quality Fits Into Packaged Development. IEEE Software 10(5):85–86

Carmel E, (1997) American Hegemony in Packaged Software Trade and the Culture of Software. The Information Society 13(1): 125–142

Champy J, (1997) Packaged Systems: One Way to Force Change. Computerworld, http://www.computerworld.com, Accessed: 19 August 2002

Chau PYK, (1994) Selection of Packaged Software in Small Businesses. European Journal of Information Systems 3(4):292–302

Chau PYK, (1995) Factors Used in the Selection of Packaged Software in Small Businesses: Views of Owners and Managers. Information and Management 29(2): 71–78

Dolmetsch R, Huber T, Fleisch E, Osterle H, (1998) Accelerated SAP: 4 Case Studies. IWI-HSG – Universitat St Gallen, St. Gallen

Elliot S. Loebbecke C, (2000) Interactive, Inter-organisational Innovations in Electronic Commerce. Information Technology and People 13(1):46–66

Fan M., Stallaert J, Whinston AB, (2000) The Adoption and Design Methodologies of Component-Based Enterprise Systems. European Journal of Information Systems 9(1): 25–35

Fleck J, (1994) Learning by Trying: The Implementation of Configurational Technology. Research Policy 23(6):637–652

Friedman AL, Cornford DS, (1989) Computer Systems Development: History, Organization and Implementation. John Wiley and Sons, Chichester

Golland ML (1978) Buying or Making the Software Package That is Best for You. Journal of Systems Management 29(8):48–51

Gross PHB. Ginzberg MJ, (1984) Barriers to the Adoption of Application Software Packages. Systems, Objectives, Solutions 4(4):211–226

Hanseth O, Braa K, (1998) Technology as Traitor: Emergent SAP Infrastructure in a Global Organization. In Hirschheim, R., Newman, M. and De Gross, J. I., Proceedings of the 19th International Conference on Information Systems Association for Information Systems: Helsinki, Finland:188–196

Heikkila J,Saarinen T, Saaksjarvi M, (1991) Success of Software Packages in Small Businesses: An Exploratory Study. European Journal of Information Systems 1(3):159–169

Hislop D, Newell S, Scarbrough H, Swan J, (2000) Networks, Knowledge and Power: Decision Making, Politics and the Process of Innovation. Technology Analysis and Strategic Management 12(3):399–411

Holland CP, Light B, (1999) Global Enterprise Resource Planning Implementation. In Proceedings of the 32nd Annual Hawaii International Conference on System Sciences, IEEE Computer Society Press

Howcroft D, Light B, (2006) Reflections on Issues of Power in Packaged Software Selection. Information Systems Journal 16(3):215–235

Jones M, (1995) The Contradictions of Business Process Re-Engineering. In Burke, G. and Peppard, J., Examining Business Process Re-Engineering, The Cranfield Management School Series, London:43–59

Kautz K, Pries-Heje J, (1996) Diffusion and Adoption of Information Technology. London, Chapman & Hall

Kase SE, Trauth EM, (2003) Assimilation, Accommodation and Activism: How Women in the IT Workplace Cope. In Proceedings of the Information Resource Management Association International Conference, Philadelphia

Kieser A, (2003) Managers as Marionettes? Using Fashion Theories to Explain the Success of Consultancies. In M. Kipping and L. Engwall., Management Consulting: Emergence and Dynamics of a Knowledge Industry. Oxford University Press, Oxford:167–187

King JL, Gurbaxani V, Kraemer KL, McFarlan FW, Raman KS, Yap CS, (1994) Institutional Factors in Information Technology Innovation. Information Systems Research 5(2):139–169

Klaus H, Rosemann M, Gable GG, (2000) What is ERP?. Information Systems Frontiers 2(2):141–162

Klepper R. Hartog C, (1992) Trends in the Use and Management of Application Package Software. Information Resources Management Journal 5(4): 33–37

Kline R, Pinch T, (1999) The Social Construction of Technology. In D. Mackenzie and J. Wajcman, The Social Shaping of Technology, Open University Press, Buckingham: 113–115

KPMG (1998) Exploiting Packaged Software. KPMG, London

Li C, (1999) ERP Packages: What's Next?. Information Systems Management 16(3):31–35

Light B, (2001) The Maintenance Implications of the Customisation of ERP Software. The Journal of Software Maintenance: Research and Practice 13(6):415–430

Light B, (2003) An Alternative Theory of Legacy Information Systems. In 11th European Conference on Information Systems Naples

Light B, Holland C, Wills K. (2001) ERP and Best of Breed: A Comparative Analysis. Business Process Management Journal 7(3):216–224

Light B, Wagner E, (2006) Integration in ERP Environments: Rhetoric, Realities and Organisational Possibilities. New Technology, Work and Employment 21(3): 215–228

Lynch RK (1984) Implementing Packaged Application Software: Hidden Costs and New Challenges. Systems, Objectives, Solutions 4(4):227–234

Lyytinen K, Damsgaard J, (2001) What's Wrong with the Diffusion of Innovation Theory. In Proceedings of the IFIP Tc8 Wg8.1 Fourth Working Conference on Diffusing Software Products and Process innovations, Kluwer B.V., Deventer, The Netherlands:173–190

Markus ML, Bjørn-Andersen N, (1987) Power Over Users: Its Exercise by Systems Professionals. Communications of the ACM 30(6): 498–504

Moore GC, Benbasat I, (1991) Development of an Instrument to Measure the Perceptions of Adopting an Information Technology Innovation. Information Systems Research 2(3): 192–220

Nelson P, Richmond W, Seidmann A, (1996) Two Dimensions of Software Acquisition. Communications of the ACM 39(7): 29–35

Newell S, Huang JC, Tansley C, (2002) Social Capital in ERP Projects: The Differential Source and Effects of Bridging and Bonding,. In Proceedings of the 23rd International Conference on Information Systems Association for Information Systems, Barcelona, Spain:257–265

O'Neill HM, Pouder RW, Buchholtz AK, (1998) Patterns in the Diffusion of Strategies across Organisations: Insights from the Innovation Diffusion Literature. Academy of Management Review 23(1):98–114

Pan SL, Huang JC, Newell S, Cheung A, (2001) Knowledge Integration as a Key Problem in An ERP Implementation. In Proceedings of the 22nd International Conference on Information Systems Association for Information Systems: New Orleans, USA:321–327

Papazafeiropoulou A, (2002) A Stakeholder Approach to Electronic Commerce Diffusion. PhD thesis, Brunel University, London

Pollock N, Williams R, Procter R, (2003) Fitting Standard Software Packages to Non-Standard Organizations: The 'Biography' of an Enterprise-Wide System. Technology Analysis and Strategic Management 15(3): 317–332

Pozzebon M, (2001) Demystifying the Rhetorical Closure of ERP Packages. In Proceedings of the 22nd International Conference on Information Systems Association for Information Systems, New Orleans, USA:329–337

PriceWaterhouse, (1996) PriceWaterhouse Information Technology Review 1995/1996. PriceWaterhouse, London

Remenyi D, Sherwood-Smith M, White T, (1997) Achieving Maximum Value From Information Systems: A Process Approach. John Wiley and Sons, Chichester

Rogers EM, (1995) Diffusion of Innovations. Free Press, New York

Rogers EM, (2003) Diffusion of Innovations. Free press, New York

Ross JW, (1999) Dow Corning Corporation: Business Processes and Information Technology. Journal of Information Technology 14(3): 253–266

Russell S, Williams R, (2002) Social Shaping of Technology: Frameworks, Findings and Implications for Policy with Glossary of Social Shaping Concepts. In K. H. Sørensen and R. Williams, Shaping Technology, Guiding Policy: Concepts, Spaces and Tools. Edward Elgar, Cheltenham:37–131

Sawyer S, (2001) A Market-Based Perspective on Information Systems Development. Communications of the ACM 44(11):97–102

Scheer AW, Habermann F, (2000) Making ERP a Success. Communications of the ACM 43(4):57–61

Scott JE, Kaindl L, (2000) Enhancing Functionality in an Enterprise Software Package. Information and Management 37(3):111–122

Scott JE, Vessey I, (2000) Implementing Enterprise Resource Planning Systems: The Role of Learning from Failure. Information Systems Frontiers 2(2):213–232

Shang S, Seddon PB, (2003) A Comprehensive Framework for Assessing and Managing the Benefits of Enterprise Systems: The Business Manager's Perspective. In Shanks, G. Seddon, P.B. and Willcocks, L.P. Second-Wave Enterprise Resource Planning Systems: Implementing for Effectiveness, Cambridge University Press, Cambridge:74–101

Silva L, Backhouse J, (2003) The Circuits-of-Power Framework for Studying Power in Institutionalization of Information Systems. Journal of the Association for Information Systems 4(6):294–336

Sørensen KH, (2002) Social Shaping on the Move? On the Policy Relevance of the Social Shaping of Technology Perspective. In K. H. Sørensen and R. Williams, Shaping Technology, Guiding Policy: Concepts, Spaces and Tools. Edward Elgar, Cheltenham: 19–35

Suchman L, Bishop L, (2000) Problematizing 'Innovation' as a Critical Project. Technology Analysis and Strategic Management 12(3):327–333

Sumner M, (2000) Risk Factors in Enterprise-wide/ERP Projects. Journal of Information Technology 15(4):317–327

Swan J, Newell S, Robertson M, (2000) The Diffusion, Design and Social Shaping of Production Management Information Systems in Europe. Information Technology and People 13(1):27–45

Swanson EB, (2003) Innovating with Packaged Business Software: Towards and Assessment. In Shanks, G. Seddon, P.B. and Willcocks, L.P. Second-Wave Enterprise Resource Planning Systems: Implementing for Effectiveness, Cambridge University Press, Cambridge:56–73

Wagner E, Scott S, Galliers RD, (2006) The Creation of "Best Practice" Software: Myth, Reality and Ethics. Information and Organization 16(3):251–275

Wagner E. Newell S, (2006) Repairing ERP: Producing Social Order to Create a Working Information System. Journal of Applied Behavioral Research 42(1):40–57

Walsham G, (2002) Cross-cultural Software Production and Use: A Structurational Analysis. MIS Quarterly 26(4):359–380

Weing RP, (1984) Finding the Right Software Package. Journal of Information Systems Management 8(3):63–70

Willcocks L, Sykes R, (2000) The Role of the CIO and IT Function in ERP. Communications of the ACM 43(4):32–38

Exploring Functional Legitimacy Within Organisations:
Lessons to Be Learnt from Suchman's Typology.
The Case of the Purchasing Function and SAP Implementation

Séverine Le Loarne[1], Audrey Bécuwe[2]
[1]Grenoble Ecole de Management
[2]Ecole des Dirigeants et Créateurs d'Entreprise (EDC Paris)

7.1 Introduction: Revisiting Weber's Bureaucratic Organisation. Challenging the Power of Functions and Their Quest for Legitimacy Within Organisations

This paper aims to put forward a preliminary model for defining the legitimisation process of functions within organisations. To address this issue, we shall discuss the role of functions within organisations and explore the concept of legitimacy itself.

Weber (1921) is indirectly, the father of both concepts. He was the first to attach importance to what he called a "legitimate order" and the "legitimisation of power structures" (Ruef and Scott, 1998). He focused on the analytical issues of legitimacy that involve (1) the nature of the authority that confers legitimacy, (2) the sources of legitimacy, and (3) the processes by which legitimacy is maintained. He was also first to describe the so-called bureaucratic organisation in which actors are governed collectively by rational-legal decision rules and play a specific role related to the functions they fulfil.

Based on this paradigm, the purpose of the study featured in this article is to examine functional legitimacy and, more specifically, how a function, i.e. group of actors who fulfil the same function within an organisation, can maintain or gain legitimacy with respect to other functions within that same organisation. Despite the fact that research on the topic remains rather scarce, we assume that functions employ several means to earn legitimacy with respect to other functions, broaden their scope of action, and indeed, develop a specific form of domination over those functions. This topic has already been developed and discussed for specific functions such as Finance (Fliegstein, 1987), Human Resources (Murphy and Southey, 2003), and, more indirectly, Marketing functions (Keith, 1960; Levitt, 1960; Tapp and Hughes, 2004). However, even though these studies highlight the legitimacy gained by these functions, that legitimacy is merely described, but not defined. This paper therefore aims to offer a first definition of the legitimacy of functions.

Since its birth, the concept of legitimacy has grown. One observes that organisational legitimacy – i.e. how an organisation can be adopted, accepted, or, at least, can survive within a specific environment, and indeed, be accepted by a community, stakeholders or society as a whole – has received most attention. Inversely, and without claiming an exhaustive literature review on the subject, it would appear that legitimacy inside the organisation – the legitimacy of the individual, the actor or the group – remains under-explored. As regards organisational legitimacy, we share Suchman's view that the concept of legitimacy can be explored from two different angles. The first is a strategic approach that considers legitimacy as an operational resource, developed or used by the organisation to maintain or improve its market foothold. The second perspective pertains to neo-institutional theory. Since its birth in 1977 (Christensen et al., 1997), neo-institutional theory regards legitimacy as a construct, that is taken for granted and is built by an organisational community in search of mimetism (Meyer and Scott, 1983). However, we presume that there is also a third approach, an organisational, critical and neo-Weberian approach that, like the neo-institutional perspective, confirms the existence of legitimacy, but regards it as a means of dominating other actors (Courpasson, 2000).

In light of these approaches and amidst all the research that has been conducted in or around the topic, Suchman's contribution to the definition of the concept of legitimacy (Suchman, 1995) is of considerable interest (Ruef and Scott, 1998). First of all, the author tries to offer a literature review that gives a clearer understanding of the concept, viewed from two different angles – strategic and neo-institutional. Consequently, he strives not only to describe organisational legitimacy within a specific context, but to define the notion of legitimacy (Suchman, 1995) by combining approaches previously developed. The second reason that "legitimises" the interest for Suchman's work is the fact that his typology and definition of legitimacy have been widely quoted and re-used in organisational studies, as well as in other areas of management science, especially in marketing (in international journals in English, see, for instance, Handelman and Arnold, 1999; Wathne and Heide, 2000; Grewal et al., 2001; Grewal and Dharwadkar, 2002; in the French Research Community, see Capelli and Sabadie, 2005).

This paper aims to test the robustness of Suchman's typology on a specific research topic – the functions within an organisation and, more precisely, the purchasing function.

To this effect, we will begin by presenting Suchman's contribution to the understanding of the concept of legitimacy as a whole, and, more specifically, the concept of organisational legitimacy. Secondly, we will test Suchman's typology on a specific case, the implementation of the purchasing module of the SAP software within an organisation. Finally, we will conclude this article by putting forward a series of proposals on how a purchasing function within an organisation can reinforce legitimacy using Suchman's typology. We conclude by raising the question of the relevance of using Suchman's typology to analyse functional legitimacy.

7.2 Suchman's Contribution and Suchman's Typology in the Debate on the Nature of Legitimacy

Before presenting Suchman's contribution to the understanding of the concept of legitimacy, we shall begin by recalling the theoretical framework he uses, and then explain how he defines legitimacy. From a theoretical point of view, the typology of legitimacy and the structure of legitimisation may explain organisational and departmental/functional legitimacies as well. However, we should bear in mind that organisational typology and organisational strategies for legitimisation remain to be empirically tested. And likewise, the strategies that departments and functions can adopt need not only be tested, but defined as well.

7.2.1 Suchman's Theoretical Framework and his Definition of Legitimacy

Suchman defines himself as a sociologist of law. More precisely, his research focuses on the impact of law firms in the structuration of the Silicon Valley (Suchman, 1994) and more generally, the impact of law on the construction of organisations and industries (Edelman and Suchman, 1997; Suchman and Edelman, 1997). Indeed, he challenges the means and ways used by law firms to gain legitimacy in this geographical area (Suchman, 1995b).

Using this study as his starting point, Suchman then broadens his perspective to focus on the process by which (any) external audiences grant some degree of approval to (any) organisations, and on the implications of different types of legitimacy on organisational activity. The article he published in the Academy of Management Review presents his approach and his definition of legitimacy (Suchman, 1995). Based on a literature review seen from a strategic and a neo-institutional standpoint, he highlights the incoherencies between the various definitions and the conception of legitimacy, and presents an inclusive definition of the concept. However, using an existing typology of legitimacy, he proceeds to develop a strategic model and explain how organisations can use that model to gain, maintain or repair legitimacy within their own environment. In doing so, Suchman sheds light on the unexpected consequences of the manipulation of legitimacy by organisations.

Our objective here is not to summarise Suchman's article, which indeed, would involve the risk of diluting its content. On the contrary, our aim is to emphasise key elements pertaining to the strategies that organisation can adopt for maintaining legitimacy in their environment. More precisely, we will highlight (1) Suchman's definition of legitimacy, which is widely quoted in other research articles, (2) his contribution to the understanding of the typology of legitimacy, as devised by Aldrich and Fiol (1994), and (3) the typology of strategies that organisations can develop to gain, maintain and repair legitimacy.

7.2.1.1 Suchman's Contribution to the Definition of Legitimacy: the Development of a Broad an Inclusive Definition

Suchman observes that many definitions have been put forward to describe the concept of legitimacy. They all share one common characteristic: they identify a correlation between the unspoken recognition of something – the organisation, etc. – by a community and socio-cultural values that are shared by each member of that community. In order to circumscribe all of the definitions given to the concept, Suchman defines legitimacy as follows:

Definition 1. "Legitimacy is a generalised perception or assumption that the actions of an entity are desirable, proper, or appropriate within some socially constructed system of norms, values, beliefs, and definitions" (Suchman, 1995: 574).

7.2.1.2 Suchman's Contribution to Aldrich and Fiol's Typology: a Theoretical Construct for a Better Understanding of the Concept of Legitimacy

Much of the literature published on the topic of legitimacy post-1995, attributes the paternity of this typology to Suchman. However, Suchman himself acknowledged that this typology was developed by Aldrich and Fiol (1994). In fact, his literature review substantiates the typology put forward by Aldrich and Fiol. The literature reveals three types of legitimacy that external audiences may grant to an organisation: an interest-based pragmatic legitimacy, a value-oriented moral legitimacy, and a culturally-focused cognitive legitimacy. These are summarised in Table 7.1 and discussed below.

Table 7.1. Types of organisational legitimacy by Suchman (1995)

Type	Definition
Normative legitimacy	Organisation reflects acceptable and desirable norms, standards, and values.
Pragmatic legitimacy	Organisation fulfils needs and interests of its stakeholders and constituents.
Cognitive legitimacy	Organisation pursues goals and activities that fit with broad social understandings of what is appropriate, proper, and desirable.

Even though Suchman did not invent this typology, his main contribution lies in his capacity to bring together other typologies or elements that have already been characterised on legitimacy. He succeeds in gathering all aspects of legitimacy into a 10 block matrix, depending on the position of each characteristic with respect to two main categories. The first category refers to the origin of legitimacy: whether it comes from the organisation – and a conscious effort made by the organisation – or by "essence". The second category concerns a legitimisation process that can be episodic or continual. The analysis of this typology through these two axes is most interesting and is based on the following assumption: organisational legitimacy is not just a passive statement, which can be regarded as the result of interactions between organisations. On the contrary, organisational legitimacy can also be considered as the outcome of strategies developed and adopted by organisations.

However, it is rather a tricky task to distinguish what aspect of legitimacy is based on a "natural" fact and what other aspect is based on a firm's active strategy! The same can be said of the second axis: which criteria distinguish an "episodic" process from a "continual" one? Otherwise, one can also consider that this typology is a theoretical construct that consolidates previous typologies built to characterise legitimacy. However, this still begs the question of whether the construct is operational, i.e. whether it effectively reflects organisational reality. Indeed, the assumption that previous typologies of legitimacy are operational does not necessarily imply that a supra-typology, which includes or takes other typologies into account, is operational!

7.2.1.3 Suchman's Typology of Legitimisation Strategies

As implied in his typology, Suchman considers that there is a correlation between time, process and legitimacy. Assuming that legitimacy is not the only outcome accepted passively by an organisation, legitimisation is the process by which managers adopt different strategies to maintain, gain or repair legitimacy. Presenting a typology of all the strategies that managers can employ to reach this goal comes as a direct consequence of the development of the typology of legitimacy. However, Suchman contributes to Aldrich and Fiol's typology of legitimacy, which is already based on a neo-institutional approach, simply by adding more neo-institutional based typologies. Therefore, his principal strategic contribution resides in his typology of legitimisation strategies. While discussing the legitimisation strategies of all three groups of legitimacy, Suchman argues that every possible managerial action belongs to one of those three groups, i.e. gaining, maintaining or repairing legitimacy. He brings together the possible variants in a 3×3 matrix where the rows correspond to a pragmatic–moral–cognitive "trichotomy", and the columns display a gaining–maintaining–repairing "trichotomy". This matrix is set forth in Table 7.2.

Similar to the typology of legitimacy, this typology of strategies for gaining, maintaining and repairing legitimacy is based on a literature review. No empirical validation is proposed in this article, however.

By defining (1) different types of legitimacy and (2) several axes for identifying strategies that managers can adopt to gain organisational legitimacy, Suchman's work gives an interesting twist to the analysis of the ways in which departments/functions can gain legitimacy and, more precisely, the types of legitimacy involved – cognitive, moral or pragmatic. However, there are two limitations to this approach. First, both typologies lack operational or empirical validation. Second, even though Suchman's typology of strategies of organisational legitimisation provides a useful framework for analysing department legitimisation within an organisation, the strategies that managers can use to pursue organisational legitimacy have to be transposed into strategies they can use to pursue departmental/functional legitimacy.

Table 7.2. Legitimisation strategies by Suchman (1995:600)

	Gain	Maintain	Repair
General	Conform to environment Select environment Manipulate environment	Perceive change Protect accomplishments Police operations Communicate subtly Stockpile legitimacy	Normalise Restructure Don't panic
Pragmatic	Conform to demands Respond to needs Co-opt constituents Build reputation Select markets Locate friendly audiences Recruit friendly co-optees Advertise Advertise products Advertise image	Monitor tastes Consult opinion leaders Protect exchanges Police reliability Communicate honestly Stockpile trust	Deny Create monitors
Moral	Conform to ideals Produce proper outcomes Embed in institutions Offer symbolic displays Select domains Decline goals Persuade Demonstrate success Proselytism	Monitor ethics Consult professions Protect propriety Police responsibility Communicate authoritatively Stockpile esteem	Excuse/Justify Disassociate Replace personnel Revise practices Reconfigure
Cognitive	Conform to models Mimic standards Formalise operations Professionalise operations Select labels Seek certification Institutionalise Persist Popularise new models Standardise new models	Monitor outlooks Consult doubters Protect assumptions Police simplicity Speak matter-of-factly Stockpile interconnections	Explain

These limitations can be resolved in part by analysing how Suchman's work was reused and how it contributed to the overall understanding of the concept of legitimacy.

7.2.2 The Contribution of Suchman's Typology to the Understanding of the Concept of Legitimacy

As indicated in the introduction, Suchman's definition of legitimacy has been widely used over the past decade. In order to "measure" the impact of his contribution, we observed when, how and by whom Suchman's work was quoted in other works of research. Our attention focused on articles published in the course of the past decade (1995–2005) in the most well-known and recognised journals in sociology, economics and management research: *Academy of Management Review* – in which Suchman's typology of legitimacy was published –, *Academy of Management Journal, Administrative Science Quarterly, Organisation Studies, American Journal of Sociology, American Sociology Review* and *Strategic Management Journal*. The choice of these journals was also guided by the fact that Suchman adopts a strategic and sociological – neo-institutional – approach. However, to be perfectly exhaustive, and as we mention in the introduction of this article, we noted that Suchman's work tends to be cited and used in the field of marketing research also. Table 7.3 provides a list of the 20 articles in which Suchman's article is quoted and Table A.1 in the Appendix details how his work was precisely used in those articles.

Table 7.3. Quotations of Suchman (1995) from 1995 to 2005 in six major journals

Title of journal	Number of articles which quoted Suchman
American Journal of Sociology	1
American Sociology Review	1
Administrative Science Quarterly	1
Organisation Studies	8
Academy of Management Review	9
Total	*20*

Our attention is immediately drawn towards two facts:
   The first one relates to the theoretical framework to which Suchman's work is associated. At first sight, Suchman seems to adopt a neo-Weberian approach, but without rejecting either the strategic or the neo-institutional framework. Indeed, he even tries to reconcile both perspectives. We indicated that his work had been reused by both communities over the past decade. And yet, Suchman does not appear to have succeeded in bridging the two perspectives: those who adopt the neo-institutional approach refer to Suchman as a neo-institutionalist. Those who prefer the strategic approach see Suchman's research as strategy-oriented.

The second point of interest is that Suchman's work was quite often quoted, but seldom in a manner that reflected his main contribution (Latour, 1987[9]). Thus, in the articles that compose our study sample, Suchman's work was generally quoted for his inclusive definition of organisational legitimacy. However, few articles relating to similar research studies consider Suchman's work as a basis for a better understanding of the concept of legitimacy. The works of Ruef and Scott (1998) and Barron (1998) are the two exceptions.

Moreover, the two typologies developed by Suchman are almost never mentioned and have never been empirically tested.

To resume, Suchman's contribution can be summarised into three main aspects:

(1) All of the authors share Suchman's opinion according to which organisations in search of legitimacy are rarely passive. On the contrary, they actively seek legitimisation through "achievement" strategies that make them conform to the external audience. They manipulate the external audience or inform the unaware audience of their activities. This idea is essentially developed in three articles (Ruef and Scott, 1998; Lawrence et al., 2001; Zajac and Westphal, 2004).

(2) Some authors, Ruef and Scott (1998) in particular, refined their propositions regarding legitimacy and recognised the multifaceted characteristic of legitimacy. Ruef and Scott also showed that the different sources of legitimisation are not independent from each other, but interconnected.

(3) Like Suchman, many authors assume that legitimacy can be regarded as an instrumental resource, which is necessary for the acquisition of other resources, and, finally, for the survival of the organisation (Pourder and John, 1996; Kostova and Zaheer, 1999; Zimmerman and Zeitz, 2002).

In order to better define the concept of legitimacy, Suchman strives to combine two different approaches, the neo-institutional and the strategic approach. His contribution covers three aspects: (1) a definition of legitimacy, that includes all the definitions and characteristics of the concept developed and proposed so far; (2) the development of the typology of legitimacy, as identified by Aldrich and Fiol and (3) a typology of the different strategies that can be pursued by managers to gain, maintain or repair organisational legitimacy. To date, the analysis of all the papers published in the six top-ranking journals in sociology and management in recent decades reveals that Suchman's main recognised contributions are (1) his capacity to propose an enlarged definition of legitimacy, (2) the fact that legitimacy is a multi-faceted concept, each aspect depending on the others, (3) the "new" status he gives to legitimacy, recognised to be a true resource that can be used by organisations. However, both typologies remain to be empirically validated.

---

[9] Latour (1987) notes: *"a paper may be cited by others... in a manner far from its own interests"* and even *"to support a claim which is exactly the opposite of what its author intended"* (p. 40).

7.3 Putting Suchman's Typology into Practice: an Analysis of the Legitimisation Process of a Purchasing Department During the Implementation of an ERP System

In order to build a preliminary model of legitimacy and legitimisation strategies for a function, based on Suchman's typologies, we adopted an abductive approach (Dubois and Gadde, 2002). In order to "grasp" how a function – which is often associated to a department within an organisation – can gain, maintain or regain legitimacy, a longitudinal study was conducted within a specific business unit of Pechiney, specialised in commercialising competencies and "Know-How", where the purchasing department took advantage of an ERP – Enterprise Resource Planning – implementation to regain legitimacy.

7.3.1 Presentation of the Case Study

7.3.1.1 Research Method and Object of Analysis
The following pages present the main results of our study that was conducted in 2001 (Le Loarne, 2001, 2005) just before, during and after the ERP implementation. The results are based on information collected from three different sources and a triangulation process (Huberman and Miles, 1991): 80% of the employees from the business unit, i.e. 25 people, were interviewed face-to-face. We also met with SAP users as well as managers or people who participated in the SAP implementation workgroup. They belonged to several different services, such as human resources, finance, accounting or various sales services. Of course, all the members of the purchasing department were interviewed. Each interview lasted between one and three hours. 90% of them were recorded. 10% were not, simply because the interviewees refused the recording. However, they did validate notes that were taken during the interviews. Each member was interviewed three times – before the implementation, while Pechiney was contemplating the move towards an ERP system, during the ERP implementation and one year after the implementation. The results also came from the analysis of several internal documents, such as written working procedures or minutes from trade-union meetings, as well as from direct observations made during our visits to the firm to conduct interviews or, as a teaching assistant, to participate in training seminars. In order to rebuild the story of the ERP implementation, the content of some discourses was compared to direct observations and, moreover, to the content of internal documents.

Before presenting the results, we should acknowledge that this single case does not allow us to apply a complete model for legitimacy and the legitimisation process developed by Suchman. It simply and only provides basic elements that will help purchasing managers identify strategies that they can adopt to gain legitimacy within the organisation.

The case we have chosen to develop here is just an episode of everyday company life. And yet, analysing the strategies developed by a purchasing department within an organisation and, more precisely, how it can benefit from the

ERP implementation to regain legitimacy, implies that managers or functions can indeed gain, maintain or regain legitimacy thanks to not only a continual, but also an episodic, process. However, this assumption remains to be validated.

7.3.1.2 Spotlight on the Context and the Reasons fo Implementing an ERP System

In 2000, the CEO and top management of Pechiney decided to implement an ERP system for two main reasons. First, the existing information system, and especially the purchasing module, which we will present in more detail later, was ageing. Secondly, the CEO wanted to pursue a cost-savings project which was launched towards the beginning of the 1990s. As leader of the ERP market, SAP was chosen. SAP was implemented in a rather traditional manner: a project group, designated by the CIO and assisted by external consultants, "re-engineered the business processes" (Hammer and Champy, 1993), established the procedures of each module and defined the parameters.

7.3.1.3 The Purchasing Function, the Purchasing Department and Purchasing Procedures Prior to the ERP Implementation

Before the ERP implementation, the purchasing process was rather informal, and almost anyone could order what they thought best for the situation. Thus, if a service needed a specific material, which was not available on stock, the assistant would call the supplier and order the required quantity. As the head of the business information service, in charge of economic intelligence and procurement for the group commented: "With the old information system, we received invoices. They were stamped, signed and sent to the accounting service. Then, once the order was received, a fax was sent and the invoice was paid''.

In this context, the purchasing department had been created a few years earlier, following recommendations expressed by the head of the division. Its first mission was to select suppliers and negotiate the best prices for all materials. They achieved this mission relatively quickly, but the recommendations given by the department to the managers and actors of other departments were rarely followed. The general idea of group-purchasing materials at better prices to save money was theoretically accepted by everybody. However, many actors stressed their need for specific materials that were not included in the recommended list issued by the purchasing department.

7.3.1.4 The Quest for Legitimacy by the Purchasing Department of Pechiney: Strategies Adopted by Purchasing Managers During the ERP Implementation

Implementing a complete ERP system involves implementing several modules. Each module is related to a specific organisational function, such as accounting, finance, purchasing, human resource management, production, etc. Initially, the CIO and top-management decided to implement the ERP system throughout the whole group. The accounting and finance modules were implemented first.

However, top managers, belonging to the purchasing department within the holding, raised discussions with the financial director, the head of the information system department in charge of implementing the ERP and the executive directors to suggest that the purchasing module be implemented as soon as possible, i.e. right after the implementation of the accounting and finance modules. The justification of this request was simple: one of the strategic objectives for the whole group was to cut costs. And indeed, human resources and purchasing are the two principal sources of expenditure. So in this respect, it was logical that the purchasing department come under closer scrutiny. A few weeks later, top management announced the implementation of the purchasing module for the whole Péchiney Group.

The main idea was to implement the ERP in order to help the group reduce its spending and homogenise its working procedures worldwide. The head of the information systems department explained:

"Why did we implement SAP? To make everybody work with the same procedures and to have a clear vision, at all times, of the financial situation of the group. As for the purchasing module, we managed to make everybody work with the same buyers. For every factory, for every service, purchases are now the same"

These first comments were corroborated by the head of the purchasing department, who was also a member of the project group in charge of implementing the purchasing module:

"We cleverly used the procedures defined while implementing SAP to organise the way people purchased. We realised purchasing was the second biggest expenditure of the group. So, if everybody watched his spending, we were sure to save money".

With the SAP system, came a global procedure, which everybody had to follow. It is set forth in Figure 7.1.

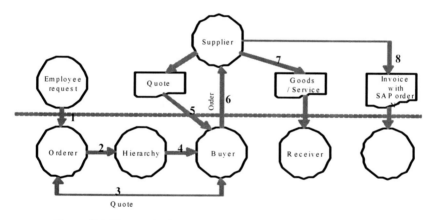

Figure 7.1. The purchasing procedure with the ERP (Péchiney, 2000)

We can note that these new procedures led to several modifications in terms of task organisation. Indeed, all SAP users from every business unit, from every country, were forced to follow a series of procedures in a specific order, which seemed to correspond to the best practices observed within the group. If they refused to do so, SAP simply did not work. An SAP user, working in the accounting service, explained:

"As you can note, you can see fields you have to fill in on the computer screen. If you don't, the transaction will fail: in this way, you are sure not to forget anything and you can follow the procedure properly".

As for the procedures, one can notice that (1) each order had to be assigned to a designated purchaser, as opposed to just any purchaser, as was the case before, and (2) the purchasing department validated the order (or not), once it had been validated by the head of the department requesting the order.

7.3.1.5 The Final Purchasing Procedure and Strategies Adopted by Purchasing Managers to Fight Resistance

The compulsory procedures were more or less accepted by employees. During our visits to the Péchiney's head-office, we noticed that the great majority of employees had to change their task organisation. That is why 70% of the people interviewed thought that SAP formalised things considerably, not because managers or other people had devised formal procedures, but just because that was how the tool worked. Indeed, to order a service or materials, they had to plug all of the required information into the software. If they did not know something, they had to go looking for it. Consequently, some employees claimed that the ERP system created too much work.

A second source confirming this resistance to change claimed that the ERP appeared to be useful for standard transaction processes, such as ordering software, supplies (…) but was too restricted when ordering complex services, such as freight. An ERP program trainer explained:

"The procedure established with SAP raises a huge problem when you have to take a decision in a hurry or if you do not have all the required information when processing an order. Indeed, let us take the specific case of shipping. Only two or three employees are concerned by this process within the whole group. The main rule with shipping transport is to work with a broker. The latter must be paid three days after the order, even if you do not exactly know all the trading conditions involved in the service. In this case, SAP does not allow us to work properly!"

These difficulties often led to procedure diversions that were developed by the purchasing department itself: Orders were often passed without being recorded on the system and were "regularised" later.

On the other hand, even though the procedure was followed on the whole by all employees who had accepted the new task organisation, a handful of managers still refused to change their habits. A trading assistant, whose boss refused the SAP implementation, said:

"At the present time, my boss doesn't even know his SAP log-in number. Everybody thinks he validates orders. In fact, he doesn't. I connect myself with his log-in number and I validate for him".

Notwithstanding the above, even if the manager forgot or refused to validate orders, the purchaser explained that if he waited for the purchase of the product or service long enough, he ended up accepting the request even though it was not appropriately or correctly formulated.

The analysis of the consequences of SAP implementation, especially the purchasing module, illustrates that this tool tends to empower purchasers who now have the ability and the legitimacy to control top management decisions in terms of purchasing. The purchaser more specifically controls and validates the process, as one of them explained:

"I spend my life controlling and checking. For instance, I negotiated prices with specific suppliers. I still find people who manage to buy things from others! In this case, I call those who issued the order, mainly managers. And, as I'm sure you know, managers hate being controlled. Thanks to procedures, SAP is THE tool that makes the process more efficient".

7.3.2 Application of Suchman's Typology and Discussion

Suchman (1995) defines legitimacy as "a generalised perception or assumption that the actions of an entity are desirable, proper, or appropriate within some socially constructed system of norms, values, beliefs, and definitions". The purchasing department existed before the ERP implementation. And, since top-management gave birth to it, we could assume, in a sense, that it had its own legitimacy. However, since its recommendations were never followed, since its actions were not recognised by the other actors of the organisation, and perhaps even by top management, we could also conclude that the department had no real legitimacy at all. So it turns out, in this case, that the ERP implementation was a means for the department to gain – and not repair – legitimacy.

Suchman's generic categorisation of the strategies available to managers in pursuit of organisational legitimacy, enables us to identify and categorise the strategies used by the purchasing department to gain legitimacy within the organisation. These strategies are summarised in Table 7.4.

So, what can we conclude from this analysis based on Suchman's categorisation of strategies for organisational legitimacy? Two main lessons can be learnt from the analysis of the strategies employed by managers of the Péchiney purchasing department in their quest for legitimacy:

All of the strategies developed by the managers of this department can be integrated into Suchman's typology. Consequently, we can consider that this typology is relevant in substantiating the content of our case study.

However, even though the strategies developed by managers can be based on three, and only three, levels, the "general" category proposed by Suchman needs to be changed: the question here is not to conform to the environment, nor indeed to select or manipulate that environment. On the contrary, the main issue would be to conform to the strategy developed for the group.

Moreover, this case reveals one aspect of legitimacy that is not covered by Suchman's typology. Indeed, managers adopt different strategies depending on who they have to convince, or who will legitimate the department. Our case reveals two main categories of people who can consider the department as legitimate: top managers on the one hand, and the rest of the organisation as a whole. We observe that managers adopt different strategies depending on who they have to convince. Managers of the purchasing department are inclined to convince top-management by playing on the pragmatic level. They tend to develop moral-based strategies to gain legitimacy from middle-management. Finally, they are more likely to adopt pragmatic and cognitive based strategies to gain legitimacy from the rest of the organisation. These observations constitute hypotheses that require validation.

Table 7.4. Comparison between strategies pursuing organisational legitimacy and strategies pursuing functional / departmental legitimacy (according to the case study)

	Identified Strategies for gaining organisational legitimacy (Suchman, 1995)	Identified Strategies pursued by managers of the purchasing department to gain legitimacy
General	Conform to environment Select environment Manipulate environment	Conform to the top-down strategy Select elements of the top-down strategy Manipulate top managers to convince them to implement the system Manipulate the project group to make them adopt the working procedure
Pragmatic	Conform to demands Respond to needs Co-opt constituents Build reputation Select markets Locate friendly audiences Recruit friendly co-optees Advertise Advertise product Advertise image	Conform to demands: - of top managers: lay emphasis on the strategy of the group (cutting purchasing costs) Select markets: - No information Advertise: - to top managers: explain the advantages of implementing a purchasing system - to users: explain the new working procedure
Moral	Conform to ideals Produce proper outcomes Embed in institutions Offer symbolic displays Select domains Decline goals Persuade Demonstrate success Proselytism	Conform to ideals: To middle-management: emphasise their need to "master" and "supervise" spending for "the future" of the organisation To users and managers: emphasise the need to save costs and share common suppliers Persuade: - Accept to change the procedure when it is too difficult to apply
Cognitive	Conform to models Mimic standards Formalise operations Professionalise operations Select labels Seek certification Institutionalise Persist Popularise new models Standardise new models	Use an tool that is: Well-known and diffused within other organisations Develop a formalised working procedure

This initial case study remains limited in terms of testing Suchman's typology of legitimisation strategies as a whole. However, it does test one part of the typology, i.e. strategies that managers can pursue for gaining legitimacy. The study reveals that strategies adopted by managers from the purchasing department of Péchiney develop moral, cognitive and pragmatic strategies to gain legitimacy not for themselves, but for their function.

And of course, this first study does not allow us to draw definitive conclusions about what kind of strategies can be adopted by managers in order to gain legitimacy. But, it does offer two contributions:

It illustrates the limitations of Suchman's typology in explaining how managers operate in order to gain departmental/functional legitimacy, and, maybe, also organisational legitimacy: our case reveals that managers tend to adapt their legitimisation strategies to the situation at hand, and, more precisely, to the person by whom they wish to be perceived as legitimate. This first result is substantiated by other research studies examining the complexity for an organisation – and not only for a department or a function – to find legitimacy. Louche (2004) lays emphasis on the multiplicity of external actors who recognise the legitimacy of the organisation. However, he states that the organisation cannot be perceived as legitimate by all these actors: *"The organisation has to choose to which norms it wants to conform and to which it does not want to conform"*.

It allows us to draw three hypotheses with regard to the nature of the process that managers may use to convince different targeted "audiences" and gain departmental legitimacy:

H1: Managers tend to convince top-management by developing strategies on a pragmatic level.

H2: Managers tend to develop moral based strategies to gain legitimacy from middle-management.

H3: Managers tend to adopt pragmatic and cognitive based strategies to gain legitimacy from the rest of the organisation.

Even though the article on organisational legitimacy published by Suchman in 1995 is often quoted, the recognition of his contribution to the understanding of the concept of organisation legitimacy remains uncertain. One major issue, among others, would be to empirically test the two typologies he developed. However, in light of this first case study, we can argue that Suchman's typology of strategies for legitimacy is relevant in the analysis and the understanding of the pursuit of legitimacy by a department or a function within the organisation. This assumption is indirectly confirmed by Ruef and Scott (1998), who use Suchman's work to build a theoretical model of organisational legitimacy. According to them, legitimisation processes operating within organisations can be considered at various different levels: (1) entire organisational populations, (2) individual organisations, or (3) sub-units and specialised aspects of organisations. However, we also argue that Suchman's work needs to be developed further and take the "target" of the legitimisation process into account.

7.4 References

Aldrich HE, Fiol CM, (1994) Fools rush in? The institutional context of industry creation. Academy of Management Review 19:645–670

Barron D.N. (1998) Pathways to Legitimacy Among Consumer Loan Providers in New-York City 1914-1934. Organization Studies 19:207–233

Beckert J, (1999) Agency, Entrepreneurs, and Institutional Change. The Role of Strategic Choice and Institutionalized Practices in Organizations. Organization Studies 20(5):777–799

Brown AD, (1997) Narcissism, identity, and legitimacy, Academy of Management. The Academy of Management Review 22(3):643–686

Capelli S, Sabadie W, (2005) La légitimité d'une communication sociétale: le rôle de l'annonceur. RAM – Recherche et Applications en Marketing PUG 20(4):53–70

Christensen S, Karnoe P, Strandgaard Pedersen J, Dobbin F, (1997) Actors and Institutions – Editor's Introduction, The American Journal Behavioral Scientist 40(4):392–396

Courpasson D, (2000) Managerial Strategies of Domination. Power in Soft Bureaucracies. Organization Studies 21(1):141–161

Dubois A, Gadde LE, (2002) Systematic combining; an Abductive Approach to case research. Journal of Business Research 55(7): 553–560

Edelman LB, Suchman MC, (1997) The legal environment of organizations. Annual Review of Sociology 23: 479–515

Fliegstein N, (1987) The intra-organizational power struggle: Rise of Finance personnel to top leadership in large corporations, 1919–1979. American Sociological Review 52(1):44–58

Grewal R, Dharwadkar R, (2002) The role of the institutional environment in marketing channels. Journal of Marketing 66(3):82–97

Grewal R, Comer JM, Mehta R, (2001) An investigation into the antecedents of organizational participation in business-to-business electronic markets. Journal of Marketing 65(3):17–33

Handelman JM, Arnold SJ, (1999) The role of marketing actions with social dimension: Appeals to the institutional environment. Journal of Marketing 66(3):33–48

Hammer M, Champy J, (1993) Reengineering the Corporation: A Manifesto for Business Revolution, Collins Business Essential, New York

Hasselbladh H, Kalinikos J, (2000) The project of rationalization: A critique and reappraisal of neo-institutionalism in organization studies. Organization Studies 21(4):697–720

Hensmans M., 2003, Social Movement Organizations: A Metaphor for Strategic Actors in Institutional Fields. Organization Studies 24(3):355–381

Huberman, AM, Miles MB, (1991) Qualitative Data Analysis. De Boeck University Press, Brussels

Jones C, (2001) Co-evolution of entrepreneurial careers, institutional rules and competitive cynamics in American film, 1895-1920. Organization Studies 22(6):911–944

Kostova T, Zaheer S, (1999) Organizational legitimacy under conditions of complexity: the case of the multinational enterprise. Academy of Management. The Academy of Management Review 24(1):64–81

Latour B, (1987) Science in Action: How to Follow Scientists and Engineers through Society. Milton Keynes: Open University Press

Lawrence TB, Winn MI, Devereaux Jennings P, (2001) The temporal dynamics of institutionalization, Academy of Management. The Academy of Management Review 26(4):624–644

Le Loarne S, (2001) L'outil de gestion, garant du pouvoir bureaucratique dans les organisations ? Etude de l'implantation d'un ERP dans une entreprise d'extraction et de transformation de matière. PhD thesis, University Jean Moulin Lyon 3, France

Le Loarne S, (2005), Working with ERP Systems. Is Big Brother Back? Computers in Industry 56(6):523–528

Levitt T, (1960) Marketing myopia. Harvard Business Review 38:24–47

Louche C, (2004) Ethical Investment. Processes and mechanisms of institutionalisation in the Netherlands, 1990–2002, Unpublished PhD Dissertation, University of Rotterdam.

Mazza C, Alvarez JL, (2000) Haute Couture and Prêt-à-Porter: The Popular Press and the Diffusion of Management Practice. Organization Studies 21(3):567–588

Meyer JW, Scott WR, (1983) Centralization and the legitimacy problems of local government. In J.W. Meyer and W.R. Scott (Eds) Organizational Environments: Ritual and Rationality, Sage Publications, Beverly Hills:199–215

McKinley W, Zhao J, Garrett Rust K, (2000) A socio-cognitive interpretation of organizational downsizing. Academy of Management. The Academy of Management Review 25(1):227–243

Mone MA, McKinley W, Barker VL, (1998) Organizational decline and innovation: a contingency framework. Academy of Management. The Academy of Management Review 23(1):115–132

Murphy G, Southey G, (2003) High performance work practices. Perceived determinants of adoption and the role of the HR practitioner. Personnel Review 32(1):73–92

Phillips N., Lawrence T.B., Hardy C. (2004) Discourse and Institutions. Academy of Management Review 29(4):635–652

Pourder R, John CHS, (1996) Hot Spots and Blind Spots: Geographical Clusters of Firms and Innovation. Academy of Management. The Academy of Management Review 21(4):1192–1225

Reed R, Lemak DJ, Hesser WA, (1997) Cleaning up after the cold war: management and social issues. Academy of Management. The Academy of Management Review (22)3:614–642

Ruef M, Scott WR, (1998) A multidimensional model of organizational legitimacy: hospital survival in changing institutional environments. Administrative Science Quaterly 43(4):877–904

Sahay S, Walsham G, (1997) Social structure and managerial agency in India. Organization Studies 18(3):415–444

Schneiberg M, Bartley T, (2001) Regulating American Industries: Markets, Policies, and the Institutional Determinants of Fire Insurance Regulation. The American Journal of Sociology 107(1):101–146

Scott SG, Lane VR, (2000) A stakeholder approach to organizational identity. Academy of Management. The Academy of Management Review 25(1):43–62

Suchman MC, (1994) On advice of counsel: Law firms and venture capitals funds as information intermediaries in the structuration of Silicon Valley. PhD thesis, Standford University

Suchman MC, (1995) Managing legitimacy: Strategic and institutional approaches. Academy of Management Review 20(3):571–610

Suchman MC, (1995) Localism and globalism in institutional analysis: the emergence of contractual norms in venture finance, in "The Institutional Construction of Organizations: International and Longitudinal Studies", WR Scott, S Christensen Eds. Thousand Oaks, CA: Sage:39–63

Suchman MC, Edelman LB, (1997) Legal rational myths: The new institutionalism and the law and society tradition. Law and Social Inquiry 21: 903–941

Tapp A, Hugues T, (2004) New technology and the changing role of marketing. Marketing Intelligence and Planning 22(3):284–296

Walgenbach P, (2001) The production of distrust by means of producing trust. Organization Studies 22(4):693–714

Wathne KH, Heide JB, (2000) Opportunism in Interfirm Relationships: Forms, Outcomes and Solutions. Journal of Marketing 64(4):36–51

Weber M, (1921) Economie et société, Collection Agora-Pocket, Plon, Paris.

Zajac EJ, Westphal JD, (2004) The Social Construction of Market Value: Institutionalisation and Learning Perspectives on Stock Market Reactions. American Sociological Review 69(3):433–458

Zimmerman MA, Zeitz GJ, (2002) Beyond survival: Acquiring new venture growth by building legitimacy. Academy of Management Review 27:414–431

8

How to Take into Account the Intuitive Behaviour of the Organisations in the ERP?

François Marcotte
FM Consulting

Enterprises that implement ERP systems aim at controlling their global performances through formalisation and standardisation of their processes, using tools dedicated to information processing and to exchanges and communication between actors. The results are a huge amount of information available in the organisation.

We have seen from our experience that this large amount of information suggests to the actors new interpretations, new intuitive processes, none formalised in the ERP but often efficient regarding the expected performance of the organisation. Indeed, the more the enterprise is subject to uncertainty, the more it uses intuitive behaviour through less defined processes able to manage a fuzzy environment.

Our basic assumption is that the implementation of ERP systems can favour the appearance of intuitive processes, which help the organisation in managing uncertainty, while the purpose of ERP is rather to formalise and standardise the processes.

The task is then to take into account in the ERP these intuitive processes which exist around the ERP. To cope with the lack of description of the intuitive process itself we propose various concepts to describe the necessary elements required for these processes regarding the organisation perspectives. It will then be possible to integrate these elements in the ERP system.

8.1 The Enterprise: a Complex Mix of Various Trades Organised in Business Processes

Companies are nowadays facing an unstable environment with reduced visibility of their market, but have to be more and more efficient in order to satisfy tighter and tighter consumer constraints. Moreover, job complexity is growing with the complexity of products, services and technology, combined with new constraints, such as energy market strain, or environmental and health care issues. Manufacturing complex products or providing high added value services requires more and more accurate competences shared between various managers and operational actors, leading to the necessity for knowledge and information integration.

To better understand these integration requirements, it is relevant to analyse the organisation from the flow point of view. The trades, actors, functions, and resources of an organisation are used and connected thanks to the flows of information and/or flows of products. For example, following the processing of a customer order through the sales, the manufacturing planning, the supply department, the workshop, and so on, allows an understanding of the transversal use of the company resources towards the company business goals. To reach these business goals, the enterprise has identified a set of organised actions to be made; the business processes are born, with their definitions and set of related methodologies and approaches.

Usually, these business processes are defined as collections of activities designed to produce a specific output for a particular customer. Depending on the perimeter, the customer can be another process in the organisation or an external customer (see Figure 8.1 for a common business process representation).

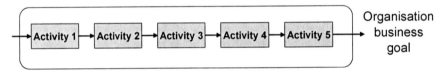

Figure 8.1. Business process representation

Business process definition implies a strong emphasis on how the work is done within the organisation. A process is thus a specific ordering of work activities across time and place, with a beginning, an end, and clearly defined inputs and outputs. In other words, this is a structure for action. An output of one business process may feed another process, either as a requested item or as a trigger to initiate new activities.

A business process has a well defined goal, which is the reason why the organisation performs this work, and it should be defined in terms of the benefits this process brings to the organisation as a whole and in satisfying the business needs.

These business processes use information to perform their activities. Information, unlike resources, is not consumed in the process – rather, it is used as part of the transformation process. Information may come from external sources, from customers, from internal organisational units and may even be the product of other processes.

8.2 Enterprise Resource Planning to Support Business Processes

To be able to easily reach their business goals, enterprises aim at formalising and standardising their business processes, through implementation of ERP systems. Such systems, dedicated to information processing and to exchanges and communication between actors, are supposed to provide the organisation with clarity of purpose and efficiency (see Chapter 5 of this book by James Taylor and Sandrine Virgil). The results are a huge amount of information processing available in the organisation, so that each activity of each business process should be performed efficiently. In fact, as long as the information processing and the communication protocols are deterministic, the ERP system provides the organisation with powerful processing capacities and communication support.

On the other hand, when the information processing is not deterministic, such as a decision making process, aggregation process, or even data analysis, the system gives control to the actors. For example, managers analyse the Sales and Operations Plans from the ERP calculation program and decide on the acceptance or not of the solution proposed by the ERP. Based on computerised demand management analysis, the sales manager will decide on the next business objectives to be assigned to his sales team. According to his knowledge of the current situation, the workshop manager will decide on the release of work orders proposed by the ERP, and on the priorities.

Thus, as long as the rules and the information to be taken into account for decision making process are not totally clarified and fixed, the actors remain key elements of the business processes. And the more the organisation has formalised its behaviour and its processes, the more computerisation is important and strong, and the more ERP brings efficiency. But it is clear that ERP will never cover all the business process activities of the organisation, since many information processing tasks remain non-deterministic.

Such a combination of data processing and human activity is the typical workflow description used to implement ERP in the organisation (Figure 8.2). In such projects, analysts usually describe the existing workflow, in which three basic types of activities are identified. First, the activities that can be supported by the standard version of the ERP are isolated. They correspond to standard transactions, like accounting rules, material requirements calculation techniques, or order release process.

Second, the activities that may be computerised are listed. They represent the gap between the ERP standard processes and the way the organisation performs these processes. Then, organisation changes or development of new ERP processes are subject to decision.

Some tasks are identified as specifically human driven activities, like decision making processes. For these tasks, all the required data available in the ERP are available to the actor.

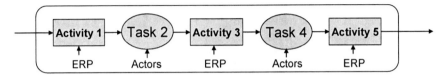

Figure 8.2. A business process described as a workflow

Finally, through the participation of actors, the business processes use various skills and know-how available within or even outside the organisation. Usually, the skills and competences are grouped in department, services or functions, often identified in the organisation as communities of actors having homogeneous trades. Following our previous example of customer order processing, one can identify the sales department, the supplying and purchasing department, the manufacturing department, and so on. This is the interesting cross-functional view of the organisation brought by the business process description (Figure 8.3).

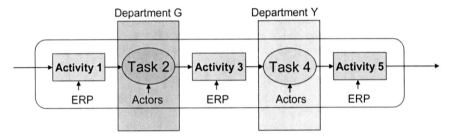

Figure 8.3. Simplified view crossing the business process and the organisation

8.3 EDME Company: a Real Industrial Example

The following industrial example, based on a real case, does not pretend to be exhaustive, but aims at illustrating our statements.

This study was conducted in the south-west part of France, where a number of companies work in the aeronautic/aerospace industry. Some specificities compared with other large industrial sectors (e.g. the electronic, agro-food or automotive industry), is that the considered products are expensive, complex to manufacture and require long cycle times. The final demand does not involve very large quantities, and is mainly subject to slow variations through time.

The typical company we consider is specialised in the machining of precision parts, and in the whole supply process towards the final aircraft, it operates as a supplier to the final assembler. Our company, that we name EDME, manufactures various parts from different aircraft programmes, and then has to manage an important product mix.

EDME has a classical organisation of 400 persons, with 7 main departments, namely Sales, Engineering, Logistic, Manufacturing, Purchasing and Supply, Administration and Accounting.

To maintain and to increase its market position, EDME makes efforts to meet delivery requirements and to decrease its cycle time, following basically a make-to-order process.

For long term coordination, EDME establishes each month a Sales and Operations Planning, defining the global production volume, based on its own capacity but taking into account the capacity of key sub-contractors. On the basis of this overall planning, business volume, costs and delivery requirements are negotiated twice a year with these direct sub-contractors, to provide flexibility for medium and short term production and inventory management. Also, this 2-year horizon plan is used by the purchasing department to organise supply.

To clarify how activities and responsibilities are shared within the organisation and to optimise the use of the existing ERP, the company made a description of its business processes. The two business processes we take as examples are the demand management process, for mid-term planning, and the customer order process.

Simplified versions of these business processes are given in Tables 8.1 and 8.2, where the decisional activities are presented in italic. While simplified, the data correspond to reality. The actors are identified by department.

In the first process, i.e. the demand management process, four main departments are involved: Sales Administration, Logistic, Manufacturing, and Purchasing, to which the supply manager belongs. Three main decisions are part of the process:

- To validate forecasts: from the customer forecast, the product manager decides on the forecasts to be integrated in the ERP for Material Requirements Planning (MRP) and workload calculation.
- To validate mid-term production plan: from the calculated workload, the manufacturing manager decides on the work allocation (internal or external).
- To validate vendor schedule: from the material requirements calculation, the supply manager allocates requirements to vendors, in the frame of the purchasing agreements.

For the second business process, i.e. the customer order process, the departments involved are the Sales Administration, the Logistic department through Product Manager and Releaser, and the Purchasing department with the Sub-contractor supplier. The business process is described in Figure 8.2. In this process, three main decisions are made:

- To make logistic acknowledgment of receipt: based on the sales and logistic agreement with the customer, the product manager decides to integrate the order in the Master Production Scheduling (MPS) Program, taking into account forecasts consumptions.
- To validate the planned orders: the purpose is to provide the scheduling department with orders to be released (Firm Planned Orders).
- To schedule Firm Planned Orders (FPO): to provide the machining centres with work programme, through FPO releasing.

Table 8.1. Company EDME Demand Management Process

Input data	Activity	Support	Who	Results	How
Paper based customer forecast plans	Input forecasts	ERP	Sales administration	Forecasts / Customer / products / Date	Using forecast input screen
Electronic customer forecasts plans	Input forecasts	ERP	Sales administration	Forecasts / Customer / products / Date	By launching EDI integration program
Internal forecasts	Input forecasts	ERP	Sales administration	Forecasts / Customer / products / Date	Using forecast input screen
Forecasts / products / date	To validate forecasts	Difference analysis table (XLS)	Product managers	Validated, modified or delayed forecasted volumes	Using gap analysis extraction (under XLS): the system extracts the previous forecasts and compares with the new one
Validated forecasts	Transfer to MRP calculation program	ERP	Product managers	Midterm load plans, Material supply requirements	By launching MPS transfer program
Mid term load plans	To validate Mid term production plan	ERP	Manufacturing manager	Sub-contracting plans (products to be sub-contracted and volume per period) Internal production plans (products and volume / period)	Select or not per product sub-contracting routings
Material supply requirements	To validate vendor forecast and order program	ERP	Supply manager	Orders and forecasts per supplier	Using Supply Plan extraction program (XLS)

These tables represent simplified definitions of the two sample business processes, with particular emphasis on the human driven activities. They were presented in the company by the logistic department manager as the various activities that should be performed during these processes, with the expected results and the supports. At this time, the level of detail was considered sufficient to specify the work to be done and the way to perform this work.

After agreement from the actors involved, implementation started and the logistic manager asked for some adjustment regarding the ERP. It took one month to implement both new ways of doing and the new ERP sub-programs.

But after some weeks of use, new discussions started between actors, particularly about clarification of the responsibilities and the real flexibility available for the decision to be made.

For example, the purchasing manager asked for sub-contracting rules as he had reduction objectives on all the external expenses, and at the same time, the manufacturing manager was deciding on sub-contracting according to the workload situation for the main machining centres.

Table 8.2. Company EDME customer order process

Input data	Activity	Support	Who	Results	How
Paper based customer orders	Integrate customer order	ERP	Sales administration	Orders / customers (product, quantity, date)	ERP order creation program
Electronic orders	Automatic integration	ERP + EDI	Sales administration	Orders / customers (product, quantity, date)	Automatic EDI integration program
Integrated customer orders with negociated prices	Make administrative A/R (Acknowledgment of Receipt)	ERP	Sales administration	Orders with administrative A/R	Using price comparison program (XLS) : the system extracts and compares the order price with the contractual price
Orders with administrative A/R	Make logistic A/R	ERP	Product Manager	Orders with validated quantity and delivery date, integrated in the Master Production Scheduling program.	Manual check from product demand program analysis (select OK in the product demand list)
Orders with administrative and logistic A/R	Make A/R to customer	ERP + EDI	Sales administration	A/R to customers	Automatic for EDI connected customer, Email or fax for others
Planned Orders (PO)	Validate PO	ERP	Releaser	Firm Planned Orders (FPO) for scheduling Purchase orders for subcontracing	Change Order status on ERP for orders to be manufacture during the next 4 weeks (horizon for scheduling)
FPO for scheduling	Scheduling	XLS	Scheduler	List of FPO to be manufactured by priority per machining centres, and allocation of the steel parts number to be used	Select the FPO among the extracted list (Available Firm Orders).
Firm orders for subcontracing	Ordering	ERP	Sub-contracting Supplier	Sub-contracting orders	Change status of Firm Order on External Order to allocated sub-contractor

Also, when the product managers were performing the logistic acknowledgment of receipt, they had no real criteria to accept or to negotiate orders with customers when these orders were different from the forecasts. So the tendency was to always accept, whatever the consequences were for the manufacturing planning and for service level.

Finally, it appeared that the first description of the business processes was not precise enough, especially to take into account a large number of different situations and possible interactions between the different departments. New requirements were raised.

8.4 Which Requirements for Business Processes in a Changing Environment?

Considering the context in which any industrial organisation operates, one can say that the changing environment directly impacts the way of reaching business goals. Facing significant environment changes, enterprises are led to perform BPR (Business Process Re-engineering, see Hammer and Champy, 1993), with more or less questioning of their current processes. In such a case, the ERP, as the business process supporting technologies, is updated simultaneously with the processes.

Besides these significant changes, most enterprises have to face continuous change of their environment, made of small steps, small modifications, small events, which do not question the business processes, but which require small adjustments of these processes, small changes in the way of doing. Such a process adjustment capability is only possible if the business processes have some autonomy in their behaviour, so that the way of doing can be adjusted according to the context, while keeping the target, which remains to reach the assigned goals. And obviously, this autonomy is only possible for non-deterministic activities, as it supposes that the activity can be performed in various ways.

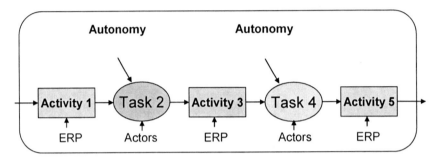

Figure 8.4. Business process with computerised activities and human decision activities

Autonomy for a process can be defined as the available degrees of freedom or decision variables regarding the way of performing the tasks required to produce the expected results (see Figure 8.4: business process with computerised activities and human decision activities). But this autonomy must be coordinated by the organisation management, because there is a risk that different actors use the same decision variable, at different stage in a business process, or in different business processes. Then, conflicting situation or incoherence may arise. For example, the decision to use overtime in a workshop should be centralised to avoid unexpected illegal situations, as there is a legal limit on overtime use. From another example, two managers deciding an inventory level should be coordinated if the inventory capacities are limited. Thus, to make this coordination, rules to use the decision variables are to be defined.

As long as degrees of freedom are allocated to the process, it will be necessary for the decision maker to build and to select one solution among the various possible ways of doing the task to reach the business process goals. And the more important the autonomy (level of freedom), the more crucial and complex will be this selection process. This is the purpose of identifying the performance objectives.

8.5 Autonomy and Competition: the Performance Weight

Regarding market competition, it is no longer sufficient for a company to be able to provide the right product or service using its well defined processes, in spite of the changing environment. It is also necessary to optimise the way it is produced, so that the overall performance remains acceptable regarding company business maintainance and development in its market. Then, with performance objectives, the business processes become more complex, in particular regarding decision processes.

As an example, production and inventory control is easier, if the costs are not taken into account; decisions to be made to synchronise material and resources availability are quite simple if over-capacity and high inventory levels are allowed. Taking into account performance optimisation, it becomes much more difficult to satisfy the customer while optimising the use of capacities and inventory levels.

So, in addition to its business goals, the organisation defines performance objectives which are then applied to the business processes, in order to control the way the business goals are reached. In particular, these performance objectives are assigned to the decision makers all along the business processes, so that arbitrages between possible solutions are made according to the expected performance optimisation (Figure 8.5).

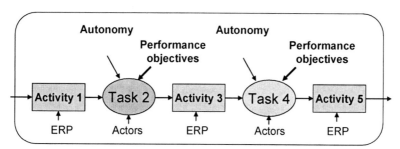

Figure 8.5. Business process with autonomy and performance objectives

These performance objectives are decomposed according to a classical top-down approach, from the global performance objectives to the local ones, so that reaching each local objective contributes to the company performance objectives. This is the purpose of coordination mechanisms: to define the rules allowing one to optimise the local performance, according to the goal to be reached, the expected performance and the available autonomy.

The goal to be reached is defined within the business processes: for example, to establish the production to be made in the next period. The expected performance will be assigned to the decision maker in the business process, such as meet the delivery dates and reduce inventory costs. Now, the available autonomy will also be assigned from the company management; for example, sub-contracting, overtime or inventory level.

So, the typical ERP workflow representative of business processes has to be combined with a structure of performance objectives, and degrees of freedom. The first one is a typical transversal view of the organisation, while the second is a typical vertical perspective, related to hierarchical delegation of responsibility (performance objectives to be reached by actors) and delegation of authority (degree of freedom given to actors in the business processes).

The actors in the organisation identify what they have to do thanks to the business process they belong to. This is the definition of their job, to perform the tasks within this business process, and they usually have been hired for these tasks, according to their skills and know-how. Then, to perform some of these tasks, actors can use ERP, in which they find modules and transactions, according to their trade.

On the other hand, the individual assessment of the actors is usually made according to the performance they reach in doing their job in the business process they belong to, according to the authority they have. This is particularly important regarding human behaviour, as this assessment according to performance objectives represents the key support for individual reward. The actor must know and understand the expectations the organisation has at him.

Thus, the business process implementation should combine two different approaches: the "ERP" view, following and supporting the work to be done to reach the business goals, and the coordination view, following the organisation to specify the performances and the autonomy for the decision makers (Figure 8.6).

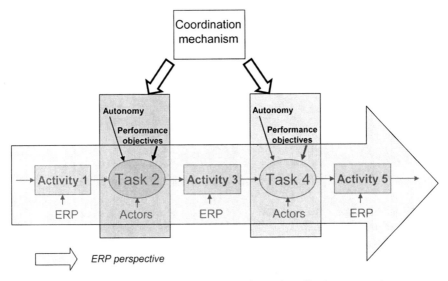

Figure 8.6. Combination of the ERP perspective and cordination perspective

Back to our example, the Logistic manager tried to clarify the various degrees of freedom allocated to decisional activities. These adjustment parameters were discussed with the actors who asked for the rules specifying the way to perform the tasks, which led the managers to clarify the expected performance for these decision processes.

8.6 Towards a Tool to Manage the Decision Processes Environment

It was clear that most of these actors had difficulty in describing and managing these elements. In fact, the decision variables and the performance were not precise enough to provide the decision maker with clear decision rules, particularly regarding interactions with other departments of the organisation. Thus, the delegation of authority was not clear.

As an example, the manufacturing manager was using many more criteria to manage the allocation of workload, internally or to the sub-contractors. He asked to have the information about the amount of the current work in progress by sub-contractor to avoid overloading them, and so to create delays. He also asked for the real sub-contracting costs, in order to follow the level of profitability for deciding on sub-contracting or not.

In releasing orders, the scheduler was basically trying to respect the FPO end dates. But he also tried to maintain 2 or 3 days of work in front of the key machining centres, to avoid any load shortage at these critical centres.

So, the logistic manager, in charge of the business process implementation and ERP improvement, asked for support, and particularly, for more theoretical supporting techniques, for modelling decision environment.

Many methods and modelling tools are available for enterprise modelling, including CIMOSA (CIMOSA Association, 1996), PERA (Williams, 1994) or PETRA (Berrah et al., 2001) which have a general purpose, others like ARIS (Scheer, 1999) being dedicated to process modelling. We have chosen here the GRAI model (Doumeingts et al., 1994) because of its well known ability to represent the decision making environment, including the elements required to coordinate the enterprise added value process, according to the performances to be reached.

Basic model: In the GRAI framework, a human decision is described through a "decision frame" which identifies the main elements required to make a decision according to the coherence requirements of the organisation (see Figure 8.7: Decision frame according to the GRAI model (Doumeingts et al., 1994)): the objectives, the decision variables and the constraints.

Objectives: they are the results or performances to be reached through the decision process. Once the performance objectives are defined, they will be structured in a hierarchy for the decision centre. Let us underline that the way this hierarchy is defined may influence the choice of a decision support method for this decision activity. A possible solution is, for instance, to consider that the main objective will be the priority (level of performance to be reached), the others becoming criteria, the optimisation of which will allow ranking possible solutions. For example, if the first objective is "customer service" and the second "cost minimisation", the manager will look for solutions which allow the added value process to respect customer requirements, and will then select the less costly solution (lexicographical approach). In that case, several optimisation criteria may be successively applied.

It is clear that the way a set of objectives is considered may influence the methods chosen for building a solution.

The performance objectives assigned to the actor is one of the results of the coordination process, which decomposes the global organisation performance objectives in local objectives towards the various actors among the various processes.

In all the cases, the performance objectives are related to performance indicators allowing monitoring their satisfaction. So from this performance objective decomposition, it is possible to specify the ERP requirements in terms of score cards or any other formatted data analysis processes.

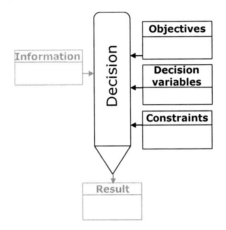

Figure 8.7. Decision frame according to the GRAI model (Doumeingts et al., 1994)

Decision Variables are parameters that modify the properties of the controlled system in order to reach the expected objectives (performances). They represent the degrees of freedom available for the decision maker. These decision variables may be local or they can be provided by other company functions or even external partners. For example, in order to meet manufacturing objectives, the planner can use local variables such as overtime, temporary workers, inventory, but could also adjust the procurement planning, in accordance with its customer represented by the commercial department, or use a network of sub-contractors via the purchasing function.

The availability of these decision variables for the actor is another result of the coordination process, which delegates authority to the actor for the use of these decision variables. Since they are supposed to help the decision maker in reaching his objectives, they should be defined in coherence with these expected performances.

Also, the use of an ERP may help the decision maker in providing simulation capabilities, like planning testing or direct costing simulation tools. So the clarification of these decision variables is important to specify the requirements regarding the ERP facilities.

Constraints are limits on the use of a decision variable. These constraints may have three origins:

- Type 1: they may express technological, contractual or legal limitations in the use of the decision variables, like "sub-contracting has to be planned two weeks in advance", "the overtimes are limited to 120h/month", "inventory cover is limited to 5 days", etc.
- Type 2: they can also come from external partners, like customers or suppliers. Examples are the inventory level limited by the customer, the delivery date (with penalty for delay or advance), the maximum amount of raw material the supplier can provide, or the capacity available from the sub-contractor.
- Type 3: they can also be the result of coordination mechanisms inside the organisation. For example, the inventory capacity available for the first workshop manager is limited to 2000 m^3 (33% of the whole capacity), and limited to 4000 m^3 for the second workshop (66% of the whole inventory capacity). As mentioned before, the maximum amount of extra hours available per workshop can be coordinated through the definition of constraints.

All additional required information allowing making decision, like follow-up information, backlog, inventory level, supplier capabilities, etc., is included in the "Information" box of Figure 8.7. Here also, the ERP provides an important support.

An example of decision frame is presented in Figure 8.8.

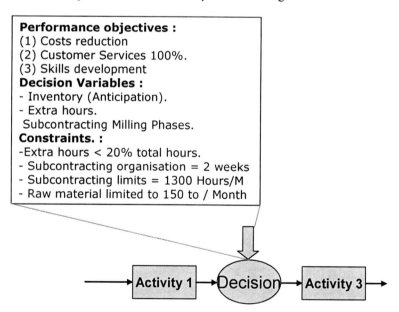

Figure 8.8. Example of decision frame

We have applied this approach in the EDME Company, starting from the business processes as they had been implemented. The initial purpose of this work was to clarify the responsibilities of the various decision makers along the business processes and to clarify their decision environment.

The first result was obviously a clear description of objectives, decision variables and constraints.

8.7 How to Transform Authority in Performance Drivers

Having a clear understanding of the performance to be reached is not sufficient for the decision maker. As long as there is no correlation between the performance objectives and the degrees of freedom, he remains inefficient in the use of his decision variables. For example, the manufacturing manager has to maintain internal workload at 100%, but should also optimise the manufacturing costs. So, in selecting the work to be sub-contracted, he was asking for information on cost levels, depending on the work to be sub-contracted and depending on the selected sub-contractor.

The question was to identify the impact of decision variables on the performance, so that he knows how to act towards the satisfaction of the assigned performance objectives. Back to the human point of view, this is the important coherence between the responsibilities and the authority. When this correspondence is not established, the person cannot perform a good job, which remains the driver to get the expected reward. Then he is subjected to the performance, instead of driving it.

The support for such a correlation is clearly the performance indicator. Bitton (1990) proposes an interesting approach to build a performance indicator system. In his work, the performance indicator is designed in relation to the decision variables and performance objectives (Figure 8.9), as the means to connect decision variable impact to performance behaviour.

We believe this is a means to consider the business process autonomy, not only as a simple adjustment variable, but rather as a performance driver.

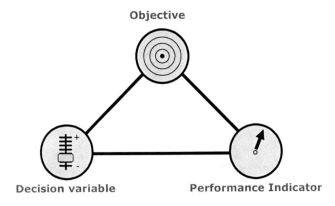

Figure 8.9. The performance management triptych

This approach was applied in the EDME company, where the performance indicators were specified according to the performance objectives and the decision variables. The results are presented in Tables 8.3 and 8.4.

This work allowed the roles and missions of each actor within the business processes to mature and to formalise. From these decision frames (objectives, decision variables and constraints), the required information has been identified, and the ERP has been modified to provide this information. Most of the calculation process and simulation tools were developed by the ERP provider, and few specific extractions towards Excel© were finally also implemented particularly to optimise performance indicators customisation, as the ERP capabilities on that subject were not the best.

All the information provided via the ERP is underlined in Tables 8.3 and 8.4. These tables represent the current version of the business processes as they are implemented.

After several weeks of utilisation, the company has improved its performance and particularly the stability of the overall production and inventory management system. The various performance indicators related to decision variables and constraints have highlighted the interactions between the decision makers. Analysing the performance change, the managers start discussions to understand and to justify these changes, and as a consequence, they also start to share their experience on the impact of their decision variables on the other decision processes.

These discussions have led to the improvement of the constraints definitions, related to the decision variables, and these constraints have allowed reduction of the amount of decision adjustments due to low coordination of these various managers' constraints.

The new ERP programs and the new distribution of access rights according to the decision frame allows one to significantly reduce the local databases created from extractions towards Excel. The company also realised the real strengths and weaknesses of its ERP, and as any imperfect ERP, many improvements remain possible, particularly concerning the customisation of interfaces.

8.8 How to Take into Account the Intuitive Behaviour of the Organisations in the ERP?

Thus, in a changing environment, with a high level of uncertainty, actors having autonomy are often spending more and more energy in trying to develop intuitive processes, analysis and interpretations, in order to increase their knowledge of the situations and the confidence in the decisions to be made, regarding the performance to be reached. The more these intuitive processes exist, the more complex is the implementation of the ERP, as it is difficult for the actors to describe a priori the standard information processing requirements. On the other hand, these intuitive processes around the autonomy provide the organisation with real adaptation capabilities which remain absolutely necessary today.

Table 8.3. Decision frame in the demand management process

Input data	Activity	Results	How	Performance objectives	Performance indicators	Decision variables	Constraints
Forecasts / products / date	To validate forecasts	Validated or modified forecasted	Using gap analysis (the new forecasts is compared to the previous one)	Reduce undeliveries due to late forecasts changes; Minimise forecast changes over the agreed conditions; Minimise finished product inventory	Reliability of forecasts inside the manufacturing cycle. # of forecast changes over the agreed conditions; Finished product inventory / customer	Modify forecasts (under customer contractual conditions). Negotiate with customer.	Contractual Conditions (with customer)
Mid-term load plans	To validate mid-term production plan	Sub-contracting plans; Internal production plans (products and volume / period)	Select or not per product sub-contracting routings	Maintain internal work load up to 100%; Maintain work load for the key subcontractors; Optimise costs (priority on classical machining process)	Internal Workload / capacity; Key subcontractors workload; Average machining costs of FPO and Order released	Adjust internal capacities for next weeks (extra hours); Adjust internal capacities for next months (night shift); Move backward or forward work loads (for delivery delay or work in process); Allocate the work load among the capable sub-contractors (according to available sub-contracting routings)	List of capable subcontractors (from purchasing department); Sub-contractors Capacities; Extra-hours limited to 10h/person/month; Night shift : 4 weeks implementation, and voluntary only; Forward workload (below 2 months); Backward workload (only with Product Manager agreement).
Material supply requirements	To validate vendor forecast and order program	Orders and forecasts per supplier	Using vendor allocation program (select capable vendor)	Favour logistic performances. Reduce costs. Favour double sourcing for aluminium	Service level per vendor; Global supply costs / month; Costs / vendors; Number of aluminium item with only one allocated vendor	Allocation to vendors (double sourcing); Increase or decrease lot size	Purchasing agreements (minimum volume for main supplier); Supplying parameters (quantity discount, minimum batch).

Table 8.4. Decision frame in the customer order management process

Input data	Activity	Results	How	Performance objectives	Performance indicators	Decision variables	Constraints
Orders with administrative A/R	Make logistic A/R	Orders with validated quantity and delivery date, integrated in the Master Production Scheduling program	Manual check from product demand program analysis (select OK in the product demand list)	100% service level. Respect ordering conditions	Service level / week Number of orders where order delivery cycle < contractual delivery cycle)	Accept or not; Negociate new dates or quantity with customer	Material availability; Level of work in progress
Planned Orders (PO)	Validate PO	Firm Planned Orders (FPO) for scheduling Purchase orders for subcontracing	Change Order status on ERP for orders to be manufacture during the next 4 weeks (horizon for scheduling)	No firm orders with release date up to 1 month Maintain projected inventory below 2 months	FPO with release date up to 1 month Projected inventory	Validate or not, Adjust PO quantity, Lot sizing	Material and tool availability; Technical data availability Lot size
FPO for scheduling	Scheduling	List of FPO to be manufacture by priority per machining centres, and allocation of the steel parts number to be used	Select the FPO among the extracted list (Available Firm Orders XLS).	Respect FPO latest end date Provide 2 to 4 days work load for each machining cell	Late Orders Workload / machining cell/ day	Order priorities (release date)	Available firm planned orders

A first illustration of this phenomenon is the huge amount of data analysis which is made with Excel© from Microsoft Office, using the table functionalities and parallel databases created by actors. And the local databases used to support analysis lead to problems of data integrity and coherence.

These capabilities of easily extracting data are now a well known key sale argument by ERP sellers! Another illustration of these trends, are the new functionalities, like demand management support tools, or other executive information systems, which are developed and implemented by ERP editors to provide more and mode information processing capabilities to the decision makers, in addition to the classical ERP functionalities. But these capabilities remain general and independent from the local decision frame.

We can see in our example that the clarification of performance objectives and decision variables has also led to the development of new facilities in the ERP. Of course, it has also led to the construction of few extractions from the ERP database, to provide the decision maker with customisation facilities which were not availale in their ERP. But these extractions were defined according to the requirements of the decision process, and they are mastered by the Information System Managers.

So, instead of trying to describe the decision process itself as a standard data processing function, we believe it is more relevant to analyse the decision context and the decision elements around this process, and to provide the relevant data to describe and understand this context. The decision process, as the way to combine the possible adjustment parameters according to the performance to be reached, will be different for each situation and will depend also on individual skills, experience and behaviours. This is a part of the intuitive behaviour of the organisation, because the way to process data and to make the decision will be context dependent. And, on the other hand, the performance to be reached and the available autonomy remain rather stable, as they are related to the organisation, defined as a set of individuals using a communication system and organised to reach goals.

Thus, to succeed in ERP implementation and use, it is important first to describe the business processes of the organisation, and to identify among these processes which are the decisional activities.

Second, it is relevant to understand the decision frame, which specifies the decisional environment, i.e. the performances to be reached, the available degrees of freedom and the related constraints.

In combining both the business process approach and the coordination view, we believe it is possible to better integrate the intuitive behaviour of the organisation with the ERP capabilities.

8.9 References

Berrah L, Clivillé V, Harzallah M, Haurat A, Vernadat F, (2001) PETRA: Un guide méthodologique pour une démarche de réorganisation industrielle. Activity Report of LGPIM, December.

Bitton M, (1990) ECOGRAI: Méthode de conception et d'implantation de systèmes de mesure de performances pour organisations industrielles. PhD thesis, University of Bordeaux 1, September

CIMOSA Association, (1996) CIMOSA Technical Baseline, Version 1.6, CIMOSA Association, e.v., D-71034, Böblingen, Germany, April

Doumeingts G, Chen D, Vallespir B, Fénié P, (1994) GRAI Integrated methodology (GIM) and its evolutions: a methodology to design and specify advanced manufacturing systems. IFIP Transactions B: Computers Applications in Technology B-14:101–117

Hammer M, Champy J, (1993) Reengineering the Corporation: A Manifesto for Business Revolution, Collins Business Essential, New York

Scheer (1999) ARIS Toolset Documentation. Saarbrücken, Germany

Williams TJ, (1994) The Purdue Enterprise Reference Architecture. Computers in Industry, 24(2-3):141–15

Process Alignment or ERP Customisation: Is There a Unique Answer?

Bernard Grabot
University of Toulouse, ENIT, LGP

9.1 Introduction

The business process re-engineering phase is recognised as a crucial step of an ERP implementation, supposed to make possible the mapping between the company activity and the ERP standard processes. Implementing the standard processes included in an ERP is indeed considered in the literature as a major condition of success of the implementation. Moreover, these standard processes, defined after multiple implementations in various sectors, are often seen as "best practices" allowing one to increase company performance, and providing a powerful tool for change management. On the other hand, many authors point out the difficulties of adoption of such external processes, and the question of knowing whether standard processes may lead to a competitive advantage is becoming widely addressed. In spite of its inconveniences, customisation of ERP package is considered by many authors as a possible solution to both adoption and competitive advantage issues.

Even if generalisation is hazardous in this field, this "ideal reasoning" can be far from industrial reality. In this chapter, we shall try to emphasise the difficulties and possible inconsistencies of each step of two opposite ways of reasoning, which consider business process alignment as a necessity, or customisation as the only chance to make an ERP package both adoptable and bring a competitive advantage.

For that purpose, we shall first try to distinguish between two interrelated but separated problems: the problem of adoption of the changes induced by the ERP seen as a tool for automating business processes (Section 9.2), then the problem of adopting external processes as provided by the "best practices" included in the ERP packages (Section 9.3). In the two cases, we shall try to compare views from the literature with industrial experiences. The issue of customisation, seen as a way to cope with the problem of appropriation of external processes, is discussed in

Section 9.4. The question of knowing whether ERP systems may lead to competitive advantages is discussed in Section 9.5.

9.2 The ERP as a Tool for Change Management

Motwani et al. (2005) suggest that the distinction between management of change and process management is a condition of success of the ERP implementation. This distinction is in our opinion not so clear, ERP implementation being an opportunity for change management, often considered as the most important task in re-engineering (Mumford and Beekma, 1994, Bruss and Roos, 1993).

Successful re-engineering, success of the ERP implementation and organisational benefits are closely linked (Law et al., 2007). Process re-engineering, which can of course be performed outside any ERP project, is indeed a major source for performance improvement. Nevertheless, the context of an ERP implementation provides both the opportunity and the tool to make change operational (Al-Mashari, 2001). In spite of this, an ERP system is only an enabler for O'Neill and Sohal (1998), who argue that 70% of re-engineering programs fail because they have been used as a substitute for strategic thinking. The concept of "enabler" is also present in Eihe and Madsen (2005), who consider that an ERP should transform the company into a more efficient and effective organisation. For some authors, ERP projects may be considered as organisational learning processes whereby the actors discover the reality and complexity of the organisation process (Besson and Rowe, 2001) and may re-design it. In that sense, the ERP implementation does not only provide a tool for proper operation of the new system, but brings also, through re-engineering, a method for better understanding the system which has to evolve.

By definition, each change sets into question an existing, possibly stable and perhaps satisfying situation, both at the individual and organisational level. Therefore, it may arouse resistance which may have different origins. While staying in the context of an ERP implementation, we would like to insist here on four points, two being generic from any re-engineering project, and two being certainly more specific to re-engineering performed in an ERP context.

We shall first quickly discuss the link between process re-engineering, change and culture of the company. The second point is that process re-engineering has an integrated view, and privileges global instead of local performance. In that case, actors acting locally may have the feeling that the newly prescribed way of working is inconsistent. Thirdly, process engineering instrumented by an ERP requires a higher interaction with a computerised, rather rigid and complex system, which leads to competence problems. Finally, process re-engineering performed during an ERP project should result in standard processes which can be rather different from what the actors would have defined by themselves. All these problems result in the issue discussed in Section 9.4, i.e. a feeling widely spread among the actors that customisation of the ERP package is a solution to cope with adoption problems.

Rather than an exhaustive discussion on these points, we shall illustrate the first three by practical examples in the following sub-sections. The fourth point, which is the main focus of this chapter, is discussed in more details in Section 9.3.

9.2.1 Process Re-engineering, Change Management and Industrial Culture

It is now clear that in an ever-changing environment (including customers and competitors), there is no steady state for a company, which has constantly to adapt its organisation to the expected variations of its environment. As stated above, this implies abandoning former successful ways of working to adopt others, supposed to be better adapted to the future context. In the context of re-engineering as part of ERP implementation, being able to make a difference between what must be kept – even if specific – and what must be changed – even if successful – is a central problem of change management.

A trivial statement is that the possibility to create a dynamic of change depends on the capacity of a company to evolve, and results in problems of adoption of the promoted changes if this capacity has been overestimated, or has not been efficiently mobilised.

The "technical" literature on change management insists much more on acceptance at the organisational level rather than at the individual level. Although individual acceptance is a very complex and interesting problem, we shall also insist here on the "organisational" aspect of acceptance, since it is more likely to result in attempts of customisation than individual acceptance does.

In the ERP literature, the comments on organisational acceptance of change are often limited to general conditions of success of the project, like "the management level believes that the company can absorb the stress related to the effort of change" (Norris et al., 1999) or "it is necessary to understand the culture of the enterprise in terms of acceptation and capacity to change" (Bancroft et al., 1998). This last reference to "culture" is in our opinion important, but it would be an error to oppose the capacity to evolve to the culture of a company. Indeed, continuous improvement clearly belongs to the culture of the most successful companies. Nevertheless, the underlying idea of a homogeneous culture within a given company seems to be more and more set into question by the industrial reality. Indeed, in the last ten years, an increased "gap of culture" has grown within large companies between "moving managers" and "stable" ones.

A first cause is that the turn-over (internal but also external) is higher at managerial level in most large companies. The consequence is that "moving" managers may have a wider view of what is done outside, but do not always fully understand in depth the culture of their present company. Moreover, they can be poorly concerned with this culture, the company being often a step in their career.

A relatively new trend has increased this gap: following a basic idea close the principles of benchmarking (take good ideas from other industrial sectors), many managers are now recruited by large companies from rather different industrial sectors. The idea is of course to benefit from a new point of view and from drastically new methods. A good example is the case of the aeronautics industry, which has massively recruited managers from the automotive sector in order to promote increases of productivity already obtained for some years in this sector. In that case, the cultural collision between "moving" and "stable" management is no longer a side effect but is the main goal of recruiting. In extreme cases, resistance to change may become an expression of the collective attachment to the basic culture of the company. This is even truer if the change has been anticipated, i.e. has been launched in order to prevent the occurrence of problems rather than addressing already present ones. In that case, increased tensions may occur between high level managers coming from other areas, hired to promote new and demanding methods, and middle level management/operational workers not really convinced of the necessity to change in the context of success.

Some years ago, we organised a day of free exchanges on their experience between twenty regional companies which had recently implemented an ERP. A very obvious statement after these exchanges was that companies which had encountered many difficulties in the past (typically companies manufacturing low value products, with a high concurrence of emerging countries) had less problems in this implementation than more successful companies (mainly companies of the aeronautics sector, with high-tech products and dominant position). In the first cases, there was no problem in adopting change, since the necessity to "change or die" was becoming a culture. In the second case, the first concern was to keep successful habits unchanged, which of course leads to difficult problems when drastic changes are concerned.

The subject of industrial training suggested to one of our students by a workshop manager of a large company implementing an ERP is a good example of this reluctance to change: the purpose was to make an exhaustive list of information and data processing facilities available for users of the previous production management system, in order to submit this list to the ERP project manager as a pre-requisite for the new system. In the same company, the first customisation requested by the users was to modify the editions of the ERP so that they look like those provided by the previous system. In that case, it was interesting to notice that suspicion of the ERP system was shared by a large group of persons from various hierarchical levels, preventing them investing in the changes to come.

Anyway, it is certainly an illusion to think that all people can be convinced to adopt new processes and it is sometimes considered that forcing people to change their behaviour may be a necessary condition for making them change their mind, once they have verified that the new processes are efficient. This attitude considers that adoption can in some cases be a consequence and not a condition of usage and is certainly not without high risks of rejection.

9.2.2 Global Versus Local Performance

Change can provide an opportunity to improve the daily life of workers, but on the other hand, improving performance may require a company to implement changes considered as necessary even if they disturb or make more complex the daily work of some individuals. In many cases, the new definition of an individual's work may even seem to be locally inconsistent, leading to poor acceptance.

If the interest of an ERP for the organisation is in our opinion doubtless, engineers have to accept that organisational acceptance requires individual acceptance as a pre-requisite, this individual acceptance being sometimes difficult for good reasons which have to be considered and properly addressed. Social works on the subject have perhaps to make the converse route, by accepting that a system which is the cause of individual discomfort may have an interest at the organisational level, which is not only a sum of individual interests.

Indeed, processes are composed of interrelated activities, and understanding the consequences of a local activity on another – possibly geographically and functionally distant – is important for making change more acceptable, especially when the impacted workstation is not the place where the improvement will be seen.

The games which are often used for training in companies try to cope with this issue: they usually allow a simplified but more global view of a company, showing the global benefits of locally demanding methods (see the numerous games illustrating the Kanban method or the Beer Game for supply chain management). Moreover, they allow one to switch the roles of the players, leading them to adopt other points of view and therefore to better understand the consequences of their own decisions on the work of their colleagues.

9.2.3 Interaction with the ERP Package

The work load is a major stress factor at work, which can be increased by several factors. The feeling of being poorly supported by the organisation aggravates stress, leading to a feeling of poor competence for performing an activity. The introduction of a complex computerised tool like an ERP questions the competence of the actors in their daily work, especially those who are not familiar with computers. A natural reaction when such a difficult evolution is required is to reject the usefulness of the new system, rather than recognising one's difficulty in acquiring new competences (the author remembers a high level manager saying that he did not need to know how to use a computer since he was not a secretary.).

We have for instance described in Hermosillo et al. (2005) the case of the implementation of the ERP Peoplesoft® in a Mexican University. Adoption problems were in that case interpreted as linked to a poor consideration of the very different competences of various types of users. Acceptance was increased by the formalisation of levels of competences for different tasks requiring interaction with the ERP, together with the definition of the corresponding training. Moreover, the implemented processes were re-modelled in order to incorporate manual operations which were not fully described in the previous version, allowing users to better understand their work in relation with the ERP.

Stress is not only brought by the necessary evolution of competences. The increased control and traceability brought by ERP systems make it more difficult to fix human errors without referring to an authority, whereas correcting a mistake is allowed by loosely automated processes. A trial and error strategy for fixing an unusual problem is no longer possible, and users understand that they have to be good at first try even when performing rarely done activities. According to Garnier et al. (2002), the process acceleration induced by automation through ERP packages has also the potential of an anxiety-producing process up to the point that managers may question the wisdom of such conversion.

As stated previously, change management can be considered as deciding what to change and what to keep. In the case of re-engineering prior to ERP implementation, we can translate it as "where to put standard processes, and where to put specific ones?". For that purpose, customisation of the ERP system has often been considered as a way to allow the coexistence between specific and standard processes. This question is discussed in the next section.

9.3 ERP Implementation and Business Process Alignment

Best practices begin to emerge as soon as generic processes become re-engineered (O'Leary and Selfridge, 1998; Davenport, 2000). Aligning the business processes of a company with best practices is usually considered as a major source of performance improvement, but some authors also consider that best practices cannot be maintained as to provide a competitive advantage in the long term (Davenport, 2000). Theory and experiences of business process alignment will be compared in the following.

9.3.1 The Problem of Business Process Alignment

Process orientation is now universally recognised as the organisation of the company activity allowing it to cope with the work fragmentation of function-based organisations. It is interesting to remember that process orientation was first applied on material flows in the 1980s, through just-in-time then lean manufacturing principles, before being considered at the business process level. This process view is a major interest of ERP systems, process-oriented information systems allowing automation of the informational and business processes while integrating the various services and departments of a company.

Implementing an ERP package in a company is a known difficult and risky task, which has motivated a huge literature (see, for instance, a recent review in Botta-Genoulaz and Millet, 2005). In the numerous papers or books dedicated to the identification of the conditions of success or causes of failures of ERP implementation, the necessity to implement standard processes, i.e. processes defined in the ERP package, is often listed in the major conditions of success (see for instance Light, 2005; Markus and Tanis, 2000; Parr and Shanks, 2000; Bingi et al., 1999; Holland and Light, 1999), while for Willcocks and Sykes (2000), many implementation problems are linked to attempts to customise the system. The main reason is that ERP systems often appear as monolithic packages, for which modification can be hazardous (Scapens, 1998) and, in all cases, hardly maintainable (Light, 2001).

The adoption of standard processes is not only a constraint for facilitating implementation: it is also often considered as a chance. This was especially true in the 1990s, when ERP was replacing legacy systems: in Cooke and Peterson (1998) for instance, the major reason given by companies for implementing SAP R/3 was its ability to standardise company processes and systems, and it was the most widely achieved benefit. For Osterle et al. (2000), "the SAP R/3 standard facilitates information integration between the individual information systems, reduces information costs and enhances its values".

In that context, implementing an ERP is not only a matter of changing software, but of improving business processes. Instead of maintaining old procedures, companies must adapt to and learn the capabilities of the new system (Bingi et al., 1999, Holland and Light, 1999; Markus et al., 2000; Parr and Shanks, 2000).

In spite of the principles of BPR (Business Process Re-engineering), promoting drastic re-definition of processes (Hammer and Champy, 2003), the alignment of the company's processes with those of the ERP package is preferred to the implementation of broad new processes. Therefore, most of the methods promoted for ERP implementation include successive phases of business process re-definition (through "as-is" then "to-be" modelling steps), then alignment of the processes obtained with those available in the libraries of the ERP package (Bancroft et al., 1997). Choosing an ERP whose standard processes are close to those defined by the company should so decrease the gap between what is expected and what is available (Hong and Kim, 2002; Osterle et al., 2000; Chiplunkar et al., 2003; Eihe and Madsen, 2005). For Bingi et al. (1999), for instance, "organisations are advised to check carefully the degree of match between their ways of doing business and the standard practices embedded in the software package, in order to avoid a painful struggling with the software when most of its modules do not fit the business".

Reality can be poorly consistent with this idea, since the choice of an ERP is usually made at the corporate level for strategic reasons, among which are increased control and standardisation. Even if users are often not consulted about this choice, involving them in the business processes re-definition is the usual way to address the adoption problems which could result from imposing external processes on the actors of the organisation. Nevertheless, this participation is sometimes considered as purely formal: Kawalek and Wood-Harper (2002) inform for instance against "... façade of user participation, whereby management engages with the rhetoric of involvement, whilst constantly aware that the outcome has already been decided upon". For the authors, participation is a useful "tool of appeasement" enabling management to implement a global standardised software system.

Indeed, the usual theoretical framework of process alignment is hardly applicable, since the redefinition of business processes involving end users should ideally result in already defined standard processes. Process alignment should be close to selecting processes in a library. No support is usually provided for this difficult task, which would require modelling the company requirements in the same formalism as the standard processes available in the ERP package, then defining tools allowing systematic comparison of the "distance" between required and available processes (Soffer et al., 2003). Such tools are not yet fully available.

Therefore, implementing standard processes often leads to adoption problems, emphasised in the literature, with a focus on strategic, social and cultural difficulties (see for instance Yen and Sheu, 2004). A usual conclusion is that the difficulty of adoption of ERP systems leads to a high number of failures: typical figures suggested some years ago were that up to three-quarters of implementations failed in 1999 (Griffith et al., 1999). Reality was certainly more complex, since at about the same time, another study stated, after a survey on 117 companies, that 34% of the companies were satisfied and 58% "somewhat satisfied" with their implementation (McNurlin, 2001). In all cases, classical reasons suggested to explain adoption problems included the differences of interest between customer organisations, who desire unique business solutions, and ERP vendors who prefer a generic solution applicable to a broad market (Swan et al., 1999). For Adam and O'Doherty (2000) a constant trade-off exists between implementers wanting zero modification and clients wanting 100% functionality.

Even if the failures of ERP implementation in the 1990s have many other origins (Grabot, 2002), such considerations led many authors to promote customisation of the ERP software as a good way to cope with the problem of matching between tool and organisation: for Davenport (1998), there is no single "best process", therefore the customisation of ERP is necessary. For Eihe and Madsen (2005), Volkoff (1999) or Light (2005), customisation may be an answer to the misalignment between functionality of the package and requirements. On the basis of industrial examples, we shall see in the next section that alignment problems may have causes other than a poor consistency between company needs and ERP standard processes.

9.3.2 Industrial Problems Linked to Alignment

According to our experience, an important problem in the alignment of business processes is that it is implicitly based on the following assumptions:

- a company can first define its requirements, then see what is available to satisfy them;
- processes have to be defined by their actors;
- users are ready to adopt best practices if they are available.

If these assumptions are considered as true, adoption problems should be the consequence of inadequacy between what is required and what is provided. According to us, these assumptions are false in many cases.

- A company can first define its requirements, then see what is available.

Defining first the required processes, then aligning them with standard processes, as for instance described in Bancroft et al. (1998) is clearly a source of dissatisfaction, and is more consistent with the specification of a new system than with the adoption of an existing one. As stated above, the design of new processes should imperatively be framed by what is available. In most companies where we have followed re-engineering projects, this step was performed before any training of the actors on the processes available in the ERP which was already chosen. In one case only, the project manager had a good knowledge of the processes available in SAP R/3, and was permanently trying to orientate the ideas on implementable processes. This is indubitably close to Kawalek's statement on "façade user participation". In the other cases, great dissatisfaction occurred when the consultants, not involved in the re-engineering phase for costs reasons, imposed radically different processes during the alignment phase.

These considerations are, for instance, consistent with O'Leary et al. (1999), suggesting a method for implementing ERP systems where the two steps (re-engineering and alignment) are performed in parallel. In that case, it is clear that knowledge of the ERP processes is a pre-requisite for both phases, together with a method for quick selection of basic standard processes on which the actors could eventually base their re-engineering effort.

- processes have to be defined by their actors

Even if part of the business processes depend on manufacturing processes, of which the actors in the company have good experience, it can be considered that other parts depend on more "universal" principles regarding generic functions like production management, human resource management, inventory control, accounting, and so on. It is interesting to notice that the existence of such "universal management principles" is perhaps one of the boundaries between engineering approaches, aiming at defining invariant solutions to specific problems, and sociologic approaches, which do not look for solutions but for understanding problems.

Anyway, it is now clear that the use of a drastically new technology or tool modifies an activity: it is, for instance, obvious that writing with a computer is a completely different activity from writing with a pen, or that communicating through e-mails is different from sending letters. Similarly, working with an integrated system like an ERP sets new possibilities which cannot be spontaneously considered by users or legacy systems.

In that case, standard processes, which take full advantage of the new possibilities of these tools, are often better than those suggested by the local actors, since their experience usually brings them to suggest improvements of their usual way of working, fixing the weak points of their previous system with minimum change. The resulting information system is then closer to a new version of their usual software than to a completely different one. In many cases, the problem is the difficulty of creating radically new solutions, taking full advantage of a complex and partially unknown tool. We have often been involved in small companies in the phase of specification of requirements, prior to the choice of production management systems. We have always noticed that the document sent to software editors was not a real set of requirements, but the specification of a weird solution, mainly based on historical habits, with an almost complete ignorance of standard but effective production management methods like MRP (Manufacturing Resource Planning).

A basic problem of process alignment is that it often encourages the users to talk in terms of solutions, and not in terms of requirements. This sets unbearable constraints on the final alignment phase and does not allow use of the full potential of the available tools. For Garnier et al. (2002) for instance, "in the past, companies put a lot of effort into optimising business processes, then searched for a software package to support it or wrote the software themselves. Firms faced the risk of automating obsolete processes or developing marketing processes for which there were no software". We do not think that this risk belongs to the past.

- Users are ready to adopt best practices if they are available.

Adopting best practices requires of course to accept change; therefore, the points discussed in Section 9.2 on change acceptance can be applied to the adoption of best practices. Another interesting issue is that, as pointed out by many authors, process re-engineering may miss to identify the situated work-practices and practical circumstances of use whereby processes are produced (Crabtree et al., 2001). This idea is close to the one expressed in O'Leary and Selfridge (1998), arguing that best practices do not exist, since they imply a universality which is not realistic. They suggest the term of "promising practices" showing that the context of use of a process and its links with other processes should be described for allowing to fully assessing its potential interest in a given case. Indeed, many authors argue that "old" processes could include something – knowledge, know-how? – which cannot be completely replaced by standard "best practices". O'Neill and Sohal (1998) agree with the necessity to re-examine periodically how to work, but insist on the danger to ignore the "embedded knowledge accumulation over many years". This would justify customisation, discussed in Section 9.4. This idea is compared to some industrial cases in next section.

9.4 Customisation of the ERP Package

Considered by some authors as the reason for many implementation problems, and by others as the condition for good adoption, the adaptation of an ERP to specific needs is a key issue of the integration of the system in the organisation. We shall first try to be more precise on the various levels of adaptation which are possible, then discuss what can really be expected from adaptation.

9.4.1 Parameterisation, Configuration and Customisation

"Customisation" of the ERP package is considered by some authors as a synonym for "configuration", and is sometimes assimilated as a parameterisation phase: customisation is, for instance, defined as "choosing between parameters without changing the ERP code" in Hong and Kim (2002). Nevertheless, most authors introduce a graduation between the different actions allowing one to adapt an ERP to the company's needs. For most authors, this adaptation is a mandatory step of the implementation: Esteves and Pastor (2003) even consider that this adaptation is the implementation phase but underline the difference between parameterisation, which is mandatory, and real customisation, which is more risky and should be conducted only in specific cases. For Soffer et al. (2003) for instance "enhancing the system's functionality through customisation is sometimes required although not desired". For Botta-Genoulaz and Millet (2005), "specific software development/too much customisation" is considered as a trap in ERP projects by 20% of the companies considered in a survey.

To our knowledge, no author makes a distinction between parameterisation and configuration: in both cases, the idea is to set-up the parameters of a software in order to adapt it to a given context of use. Therefore, a common thinking is that the result is still a standard package which will be maintained and upgraded without dedicated effort.

In contrast, the term "customisation" is usually kept for cases when the standard package is considered as providing an unsatisfactory answer to the company's needs. Whatever the "customisation" could be (including integration of other standard software or specific developments), the result is a specific software which will require specific validation and effort for maintenance and upgrade (Light, 2001). Exceptions to these considerations can be found, like Soffer et al. (2003) in which customisation is considered as the result of configuration.

In this last paper, different levels of configuration are distinguished, called "optionality levels" by the authors:

- system configuration level, controlling options affecting the software functionality throughout the entire system;
- object level, intermediate optionality level allowing different instances of an object to be handled in different manners;
- occurrence level, which applies to a single occurrence of a process or object.

Three types of system parameters can be used at the system configuration level:

- high-level process definitions, providing preconditions to user-interface sessions (e.g. a Boolean parameter indicating whether warehouse location control is implemented in the system);
- low-level process definitions, indicating the specific algorithms and rules to be applied when performing a specific action (e.g. definition of what type of action require logging a record of change history by the system);
- process parameter definitions, providing values indicating how a specific action is performed (e.g. definition of a planning horizon).

An example of object level configuration can be a parameter indicating whether location is controlled or not in a specific warehouse (once location has been activated at the system level). The occurrence level can for instance be used for defining an option of fast approval for a given delivery.

It is worth noticing that for the authors, the processes resulting from the use of these various levels of parameterisation do not necessarily match any predefined "best-practice", which partly justifies their use of the term "customisation". Indeed, a real use of these possibilities may result in never implemented or tested processes.

As a summary, it is interesting to distinguish between:

- "first level" parameterisation/configuration, resulting in instances of standard processes in the ERP,
- "second level" parameterisation/configuration, resulting in "new" processes, i.e. processes which are far from standard ones, but are completely defined in the ERP package;
- customisation, resulting from addition of other external modules and/or specific developments.

These three levels of course induce different advantages and drawbacks:

- the first level allows one to benefit from best practices and results in a standard/maintainable software, but can eventually only partially address the organisation needs;
- the second level also results in a maintainable software, but not necessarily in the implementation of "best practices" supporting change management. On the other hand, compliance with the requirements can be better than in the previous case;
- the third level should allow more flexibility in order to address the requirements, but leads to classical problems regarding maintenance and system upgrade.

These synthetic considerations show that the technical possibility to adapt an ERP to more or less specific needs is rather important. This can be surprising when considering past experiences like the one related in Stein (1998), where the implementation of SAP R/3 was abandoned by Dell, claiming that SAP was too monolithic to be altered for changing business needs. For Scapens (1998) too, ERP packages are both flexible and inflexible: flexibility can be obtained in processing the details of individual transactions or screens, but the structural and centralised approach falls short in providing suitable functions for all business companies (Bancroft et al., 1998).

According to our experience, the configuration effort is no longer limited by technical possibilities, but mainly by the time and money required for adaptation on the one hand, and by the competence required for matching the company requirements with the system parameterisation, on the other hand. Indeed, ERP experts do not seem to have standardised competences regarding deep parameterisation of the package and it seems that an identical problem can be solved though very different means in different projects. This sets the difficult problem of the capitalisation of the configuration experiments, up to the point that in some cases, real customisation can sometimes be a simpler (if not better) solution that configuration...

9.4.2 Customisation as a Means to Adapt the System to Specific Requirements

Many papers state as an established fact that even the best ERP packages can only meet part of the organisational needs of a given company (70% in Al-Mashari, 2001), but for others (see for instance Chand et al., 2005) ERP has sufficient flexibility to integrate most of the business processes of an enterprise.

A gap analysis should help to highlight areas of deficient performance (Markus, 1988), then potential for improvement through customisation (Davenport, 1993). For Light (2005), the main reasons for customisation are the following:

- the ERP package does not include a very specific functionality (e.g. a non-standard MRP calculation),
- customisation should make some documents more appealing,
- customisation of screens could help to avoid errors when too much information is provided,
- best practices could be absent from the software in some areas (which raises the problem of software maturity),

- some "historical" processes can be difficult to change, like pricing (where change may require negotiation with customers/suppliers),
- key performance indicators could be missing and require customisation,
- customisation could help to make adoption easier,
- customisation can be a form of maintenance, or may help to cope with vendor insufficiency,
- customisation may help maintain existing ways of work perceived to be of value.

We can see that many items on this list should be in most cases attainable through configuration, and we shall focus here on customisation as a way to implement non-standard processes considered as necessary. The question is of course to know how to choose these processes.

This problem is addressed in Yen and Sheu (2004) through an interesting survey. Jacobs and Whybark (2000) suggest that centralisation of information and flexibility of production systems are two major factors which govern the adequacy of an ERP package: firms having the need for strong centralised control and little flexibility in production processes could develop and implement a single set of "best practices" within an ERP. In contrast, a strong need for flexibility and little need for centralisation should cause the company to collect different processes from various ERP systems, and to integrate the corresponding heterogeneous modules through customisation. In the examples surveyed in Yen and Sheu (2004), ERP are often considered as providing efficient but bureaucratic processes, while companies having to provide a flexible and quick answer to their customers would need more flexibility than is possible within the framework of an ERP package. This point of view is shared by many SMEs, which think that the use of standard processes can be an obstacle to reactivity.

For us, the key point is that efficient enterprise management software is supposed to create a link between the various flows of resources used by the company: especially human resource, materials, information and finance. Creating this link between accounting, manufacturing and information allows traceability in the company, with the condition that the information flow really controls the material and financial flow. In many practical cases, reactivity is obtained by by-passing the information system, through direct action on the materials or financial flows (e.g. by modifying a routing or sending a manual invoice). This type of reactivity is of course not compliant with traceability constraints, and configuring the business processes in order to allow both exception handling and traceability should be the only acceptable answer, and is most of the time possible within an ERP.

Similarly, customisation for better adoption often aims at decreasing the difference between the former and the new system (see the example of Section 9.2.1). An important question is whether there is a real added value brought by these changes, or if they are considered as necessary for the comfort of the users. The latter can of course be acceptable, but must clearly be considered as such.

Another reason for customisation can be to answer the problem of including specific knowledge in the processes, as stated at the end of Section 9.3.2. ERP packages are mainly based on procedural processes, while for Eihe and Madsen

(2005), for instance, some best practices can be "knowledge-based", like procurement. How to integrate more "knowledge" into an ERP is certainly a good reason for customisation, but may lead to problems when the strategic importance of the "knowledge" to be incorporated is overestimated.

A typical area of such misunderstanding is the coding of the articles. Old information systems were only capable of storing small amounts of data. As a consequence, in such systems, the code of an article was often the only data immediately available on a screen, and an important issue was to insert a lot of information in this code (type of part, material, way it is managed, suppliers, etc.). Understanding those codes was usually the result of long experience, and allowed the holders of such knowledge to process information quicker than other operators, giving them specific prestige in a company. When an ERP system is implemented, these specific codes are replaced by "blind" ones aiming only at distinguishing the articles through unique codes. The implementation of these codes results in a feeling of loss of knowledge and competence, especially for aged workers who are in addition those susceptible to having the most problems with the new system. Indeed, codes including information are no longer necessary in present information systems which can provide on each screen all the information attached to an article, including its designation, description, etc. Therefore, adopting these new "blind" codes has no deep impact and the ERP makes widely accessible knowledge only possessed by a limited number of experienced workers. Facing this problem, several large companies decided to re-develop the coding module of its ERP to be able to keep their former code...

Customisation seems to be linked to two main issues: improving adoption, and providing a competitive advantage. The next section discusses how to know whether standard processes could bring a competitive advantage.

9.5 Can Standard Processes or Customisation Bring a Competitive Advantage?

In the literature, specific processes are often considered as able to bring a competitive advantage, implying it cannot come from standard ones, accessible by competitors. Davenport (1998) states for instance that "an enterprise system... pushes a company toward generic processes even when customised processes may be source of competitive advantages". For him, firms could lose their source of advantage by adopting processes that are indistinguishable from their competitors. In (Davenport, 2000), the same author says that a "best practice" approach requires the organisation to re-engineer its processes to fit the software. As such, "firms implementing ERP will probably not be able to maintain ERP systems as a source of competitive advantage over time". Similarly, Sor (1999) underlines the scepticism regarding the ability of "off the shelf" ERP systems to maintain an organisational infrastructure that is different to those of the competitors. In Cooke and Peterson (1998), competitive positioning was ranked least among the benefits expected after ERP implementation, with only 28% achievement level.

These considerations seem to be based on the assumption that a competitive advantage should by definition result in a difference between a company and its competitors. Therefore, shared methods and tools could not bring such advantage. On the other hand, Hunton et al. (2003) compare business performance of adopters and non-adopters from the economic aspect and on that base, suggest that ERP adoption helps firms gain a competitive advantage. For Mabert et al. (2001) implementation managers expect the availability, quality and standardisation of data to provide a "strategic" advantage... such a "strategic advantage" also comes from cycle time compression by the automation of (marketing) processes (Garnier et al., 2002).

Competitive or strategic advantage? Going further requires perhaps being more precise on the definition of a competitive advantage. In Beard and Summer (2004), the resource-based model of competitive advantage suggested in Wernefelt (1984) is applied to the ERP. According to this framework, a competitive advantage is given by a resource or capability if positive answers are given to the following questions:

- is the resource or capability valuable?
- is it heterogeneously distributed across competing firms?
- is the resource or capability imperfectly mobile? (i.e. hardly imitable).

More recently, another question was added to that list: is the firm organised to exploit the full competitive potential of its resource capabilities? (Barney, 1999).

The first question concerns obviously the performance provided by the resource or capability, whereas the last two concern its accessibility by competitors. In the case of an ERP, the answer to the first question is clearly "yes": the benefits of ERP introduction are listed in numerous papers (see for instance Falk, 2005 or Botta-Genoulaz and Millet, 2005). Yet, the answer is "no" to the last two questions since standard processes are accessible to competitors.

Indeed, the problem of getting a competitive advantage from technology widely available is not specific to ERP systems. Concerning the use of the Internet in companies, Porter (2001) notes that this technology has a levelling effect on business practices, and reduces the ability of a company to establish an operational advantage.

In fact, as illustrated by the question added by Barney, competitive advantage should also be defined in terms of results: tools like ERP systems, but also many improvement methods widely adopted in industry, like lean manufacturing, 5S, 6 sigma etc., do have a positive impact on performance. Therefore, the question is to know whether greater achievements can be expected from these efficient but well known methods, or from specific techniques, requiring a more risky investment but capable of bringing a unique advantage. According to our experience, many companies should first follow the first path, the second one being in our opinion reserved for very specific cases. This is close to what is stated in Eihe and Madsen (2005) for small to medium size companies: for the authors, the inability of SMEs to realise competitive advantage from ERP implementation is attributable to failure to proper use technology to address change in the design and structure of an organisation.

9.6 Conclusion

Paying more attention tp exceptional cases than typical ones, and reasoning on these cases is a common temptation. The field of ERP implementation is certainly a good example of this trend, since a rather widely spread way of thinking in the domain is that the standard processes included in ERP systems are rarely adapted to the real needs of a given company. In spite of its costs and risk, customisation is as a consequence seen as the only way of preserving specific processes or activities which build the competitive advantage of a company, and of improving acceptance of the system by its users.

This assertion is of course not always false, but we do believe that much more benefit can be expected from the introduction of standard processes than from the customisation of an ERP, that most of the time, customisation results from the reluctance of the users to evolve to be able to use a different (better) system, and that when adaptation is really needed, a more precise configuration of the ERP could in many cases give the same results as customisation.

Many counter-examples can of course be found, but according to us, the most important challenges during ERP implementation concern the support for change. This support is required from operators, who can have difficulties in the daily use of an ERP, but also from lower and middle level management, who can be pushed to resistance by the pressure set by a high level management not always fully aware of the culture of the company. In order to cope with this resistance, too much emphasis is perhaps set on the ERP system itself, and not enough on the new processes to be implemented. Being able to manage separately the difficulties linked to changing the work processes and those linked to the implementation of a new information system is in our opinion a first means for making easier the adoption of a system which influences the life of the entire company. Once this is done, it will be the time for considering customisation as the way to optimise the ERP implementation, and not as a way to change the package.

9.7 References

Adam F, O'Doherty P, (2000) Lessons from enterprise resource planning implementations in Ireland - Toward smaller and shorter ERP projects. Journal of Information technology 15(4):305–316

Al-Mashari M, (2001) Process orientation through Enterprise Resource Planning (ERP): A review of critical issues. Knowledge and Process management 8(3):175–185

Bancroft N, Seip H, Spengel A, (1998) Implementing SAP R/3. Englewood Cliffs, New Jersey, Manning Publications Co. (2nd edition)

Barney JB, (1999) Gaining and sustaining competitive advantage. Addison-Wesley, Reading, MA

Beard JW, Summer M, (2004) Seeking strategic advantage in the post-net era: viewing ERP systems from the resource based perspective. Strategic Information Systems 13:129–150

Besson P, Rowe F, (2001) ERP project dynamics and enacted dialogue: perceived understanding, perceived leeway, and the nature of task-related conflicts. Data base for advances in Information 32 (4):47–66

Bingi P, Sharma M, Godla J, (1999) Critical issues affecting an ERP implementation. Information Systems Management Summer:7–14

Botta-Genoulaz V, Millet PA, (2005) A classification for better use of ERP systems. Computers in Industry 56(6):573–587

Bruss LR, Roos HY, (1993) Operations, readiness and culture: Don't reengineer without consider them. Inform 7(4):57–64

Chand D, G. Hachey G, Hunton J, Owhoso V, Vasudevan S, (2005) A Balanced Scorecard Based Framework for Assessing the Strategic Impacts of ERP Systems. Computers in Industry 56(6):558–572

Chiplunkar C, Deshmukh SD, Chattopadhyay R, (2003) Application of principles of event related open systems to business process reengineering. Computers & Industrial Reengineering 45:347–374

Cooke D, Peterson W, (1998) SAP implementation: strategies and results. Research Report 1217-98RR, The Conference Board, New York

Crabtree A, Rouncefield M, Tolmie P, (2001) There's something else missing here: BPR and the Requirement processes. Knowledge and Process Management 8(3):164–174

Davenport T, (1998) Putting the enterprise into the enterprise system. Harvard Business Review 76(4):121–131

Davenport T, (1993) Process Innovation: Re-engineering work through information technology. Harvard Business School Press, Boston, MA

Davenport T, (2000) Mission critical: recognising the promise of enterprise resource systems. Harvard University Press, Cambridge

Ehie I, Madsen M, (2005) Identifying critical issues in enterprise resource planning (ERP) implementation. Computers In Industry 56:545–557

Esteves J, Pastor J, (2003) Enterprise Resource Planning systems research: an annotated bibliography. Communications of the Association for Information Systems 7(8)1–36

Falk M, (2005) ICT-linked firm reorganisation and productivity gains. Technovation, 25(11):1229–1250

Garnier SC, Hanna JB, LaTour MS, (2002) ERP and the reengineering of industrial marketing processes – A prescriptive overview for the new-age marketing manager. Industrial Marketing Management 31:357–365

Grabot B, (2002) The Dark Side of the Moon: some lessons from difficult implementations of ERP systems. IFAC Ba'02, Barcelona, July 21-26

Griffith TL, Zammuto RF, Aiman-Smith L, (1999) Why new technologies fail? Industrial Management 41(3):29–34

Hammer M, Champy J, (2003) Reengineering the Corporation – A Manifesto for Business Revolution. Business & Economics, HarperCollins Publishers

Hermosillo Worley J, Chatha KA, Weston RH, Aguirre O, Grabot B, (2005) Implementation and optimisation of ERP Systems: A Better Integration of Processes, Roles, Knowledge and User Competences. Computers in Industry 56(6):619–638

Holland C, Light B, (1999) A critical success factors model for ERP implementation. IEEE Software May/June:30–35

Hong KK, Kim YG, (2002) The critical factor for ERP implementation: an organisational fit perspective. Information & Management 40:25–40

Hunton JE, Lippincott B, Reck JL, (2003) Enterprise resource planning systems: comparing firm performance of adopters and non-adopters. International Journal of Accounting Information Systems 4:165–184

Jacobs FR, Whybark DC, (2000) Why ERP? A primer on SAP implementation. Irwin/McGrawHill, New York

Kawalek P, Wood-Harper T, (2002) The finding of thorns: user participation in enterprise system implementation. Data base for advances in Information 33 (1):13–22

Law CCH, Ngai EWT, (2007) ERP systems adoption: an exploratory study of the organisational factors and impacts of ERP success. Information and management 44:418–435

Light B, (2001) The maintenance implication of the customisation of the ERP software. Journal of Software Maintenance: Research and Practice 13:415–429

Light B, (2005) Going beyond "misfit" as a reason for ERP package customisation. Computers in Industry 56:606–619

Mabert VA, Soni A, Venkataramanan A, (2001) Enterprise resource planning: common myths versus evolving reality. Business Horisons 44(3):69–76

Markus ML, Robey D, (1988) Information technology and organisational change: causal structure in theory and research. Management Science 34(5):583–598

Markus ML, Tanis C, (2000) The enterprise systems experience – from adoption to success. In RW Smud (Ed), Framing the domains of IT research: Glimpsing the future through the past. Pinnaflex Educational Resources, Inc. Cincinnati, OH, USA:173–207

McNurlin B, (2001) Will users of ERP stay satisfied?. Sloan Management review 42(2):13–18

Motwani J, Subramanian R, Gopalakrishna P, (2005) Criticals factors for successful ERP implementation: exploratory findings from four case studies. Computers In Industry 56(6):529:544

Mumford E, Beekma GJ, (1994) Tools for change and progress: a socio-technical approach in business process re-engineering. CG Publications, UK

Norris G, Wright I, Hurley JR, Dunleavy JR, Gibson A, (1999) SAP: An Executive's Comprehensive Guide, John Wiley and Sons

O'Leary D, Selfridge P, (1998) Knowledge Management for best practices. Intelligence Winter:13–24

O'Neill P, Sohal A, (1998) Business process reengineering: application and success – an Australian study. International Journal of Operations and Production management 18(9-10):832–864

Osterle H, Fleisch E, Alt R, (2000) Business networking. Springer, Berlin

Parr A, Shanks G, (2000) A model of ERP project implementation. Journal of Information Technology 15(4):289–303

Porter ME, (2001) Strategy and the Internet. Harvard Business review 79(3):63–78

Scapens R, (1998) SAP: Integrated information systems and the implications for management systems. Management Accounting 76(8):46–48

Soffer P, Golany B, Dori D, (2003) ERP modeling: a comprehensive approach. Information systems 28:673–690

Sor R, (1999) Management reflections in relation to enterprise wide systems. In Proceedings of AMCIS'99:229–231

Stein T, (1998) SAP Installation Scuttle – Unisource cites internal problems for $168m write-off. Information Week:34.

Swan J, Newell S, Robertson M, (1999) The illusion of "best practices" in information systems for operation management. European Journal of Information Systems 8:284–293

Volkoff O, (1999) Using the structurational model of technology to analyse an ERP implementation. In Proceedings of Academy Management'99 Conference.

Wernefelt B, (1984) A resource-based view of the firm. Strategic Management Journal 5(2):171–180

Willcocks L, Sykes R, (2000) The role of IT function. Communication of the ACM 41(4):32–39

Yen HR, Sheu C, (2004) Aligning ERP implementation with competitive priorities of manufacturing firms: an exploratory study. International Journal of Production Economics 92:207–220

10

Process Alignment Maturity in Changing Organisations

Pierre-Alain Millet, Valérie Botta-Genoulaz
INSA-Lyon, LIESP

10.1 Introduction

Companies have invested considerable resources in the implementation of Enterprise Resource Planning (ERP) systems, but the outputs are strongly dependent on the process alignment maturity because of continuous change within organisations. Commonly, the initial implementation rarely gives the expected results and the post-project phase becomes of research interest (Section 10.2). Making efficient use of such information systems is nowadays becoming a major factor for firms striving to reach their performance objectives. This is a continuous improvement process where companies learn from failure and success to acquire a "maturity" in information system management. This concerns the mapping of re-engineered processes to changing organisations, the set up of software packages and technologic hardware, but also the organisation of roles, skills and responsibilities, performance control through indicators, scorecards, sometimes called "orgware".

Based on previous investigations of the project phase (Section 10.3) and on a qualitative survey of French companies with more than 1 year of ERP use, we propose (Section 10.4) a classification approach to company positions regarding their ERP use, based on both software maturity and business alignment directions. This two-axis model is a tool to help companies to evaluate their situation and prioritise their efforts to reach the correct "level of maturity". Both axis are linked and dependent: an improvement in business alignment requires a certain level of software maturity. A maturity level is defined from three points of view (operational, process, and decisional) using "alerts" (predefined malfunctioning identified with standard checklists and overstep indicators) and is associated with correction or enhancement actions.

Reorganisation of enterprises faced with changing contexts also have major impacts and consequences on their information system. These impacts must be considered in a global methodology of continuous improvement (Section 10.5). The maturity model has to be considered in its "life cycle" to take into account disruptions like scope change or new deployment, company reorganisation, and knowledge loses. Furthermore, this maturity model can be heterogeneous in the whole organisation depending on countries, subsidiaries, etc.

Because the maturity is never equal in time and scope, a main issue of management is to understand who is concerned with a dysfunction, which skills and responsibilities are involved in the corrective action, which data of the information system has to be checked, which processes can be a cause or can be affected… This deals with dependencies between all informational and organisational entities involved: roles, skills in an organisation at the management level, information, documents and processes at the information system level, programs, forms, reports and databases at the technological level. The aim is to support the ability to "drill down and up" from an actor to the data he has to understand, from the process to the minimum scope required to change it, from a new practice supported by the information system to the actors concerned (Section 10.6).

The three dimensions – maturity, time and scope – are gathered in a "model of maturity" to help to define and organise actions of the maturity learning process.

10.2 ERP: After the Project, the Post-project

10.2.1 The "Post-project" Phase in Academic Literature

Despite the wide ERP systems base installed, academic research in this area is relatively new. Like many other new Information Technology (IT) areas, much of the initial literature on ERP was developed in the 1990s and consists of articles or case studies either in the business press or in practitioner focused journals. Since 2000, academic research accelerated with the widespread implementation of ERP systems. As indicated by Botta-Genoulaz et al. (2005) in a revue of the state of the art presented in a special issue of the international journal *Computer in Industry*, new topics are studied like organisational issues of such projects (Davenport, 1998; Bouillot, 1999; O'Donnell and David, 2000; Ross and Vitale, 2000), or cultural issues (Krumbholz and Maiden, 2001; Saint Léger et al., 2002). These authors stress the importance of the initial stages of projects to take into account cultural aspects, national characteristics, organisational strategies, decision making processes, etc.

Many studies have been made of project management methodologies, which allow clarification of the main stages of an ERP implementation project (Poston and Grabski, 2001; Boutin, 2001; Kumar et al., 2003; Deixonne, 2001; Markus and Tanis, 2000; Ross et Vitale, 2000). These methodologies relate to success factors widely discussed (Al-Mashari et al., 2003; Holland and Light, 1999; Mabert et al., 2001). Some studies concern the relationship between project success factors and post-project performance indicators or user adoption (Nicolaou, 2004; Somers and Nelson, 2004; Calisir and Calisir, 2004).

It emerges that the potential complexity of an ERP project does not only lie in the ERP system on one hand or the company on the other hand, but rather in their connection (Botta-Genoulaz, 2005). This is not limited to the implementation stage, but must consider the whole lifecycle of the information system, from the initial stages – definition of the project context including cultural and management dimensions – to the downstream stages, where the results can (or not) be achieved by the "good use" of the system.

Now, if there are many publications about project methodologies or key success factors, their efficiency to represent the ERP life cycle in the company is incomplete. Somers and Nelson (2004) studied the problem of understanding who are the key players, which activities associated with enterprise system implementations are important, and when their effect is most prevalent across the IT development stages, by questioning key players of numerous projects. Their conclusion focus beyond the adoption and acceptance stages of implementation to include both pre- and post-implementation behaviour. This appears to be particularly important for ERP systems.

We are therefore interested in the "usage" of the resulting information system, in its optimisation. By "optimisation of the information system", we understand efficient use of the available technical, human and organisational resources mobilised around the integrated information system. Boundaries between implementation and optimisation are of course fuzzy. Some evolution projects concern new implementations. Consequently, questions are various and address as well the use of existing applications as the maturity of the company to begin evolutions or new projects:

- How does an ERP implementation contribute to make the organisation more effective?
- In what way has the organisation learned from the ERP project?
- Does the company make the most of the potentials of the ERP and how do they contribute to the company results?
- Is coherence ensured between the information system, the business processes, the management rules, the procedures, and the competency and practices of the users?
- Are activity-data and master-data reliable and relevant?
- Is the ERP well positioned in terms of "information system urbanisation"?

10.2.2 The Tool and Its Use

An ERP system can be studied as a technical object, a package sold by a software editor, which comprises several components (programs, documents, databases…) that will become a computer system parametered and configured for a company. But once installed, and adapted to the company requirements, it becomes one of the components of the enterprise information system, which encompasses data, documents .. and represents a part of the knowledge of players. It is at the same time an "instance" of the standard technical object that makes every implementation a specific case, and a "deployement" of this object enhanced by company and user data.

The use of this tool cannot come down to a technological definition. Besides, if with several thousand objects, an ERP system is a complicated tool, when it is implemented in a company for business process management and used by human players, it becomes a "socio-technical" and complex system (Simon, 1996; Gilbert, 2001). The human factor is also often mentioned as an obstacle in ERP projects; a case study in a large company shows the importance of payer's attitude in the success of the project: "Employee attitudes are a key factor in determining ERP implementation success or failure. Early attitudes about ERP systems, even before these systems are implemented, shape employee views that may be difficult to change once the systems become fully operational" (Abdinnour-Helm et al., 2003). The impact of employee's behaviour will strongly influence the project and its result, i.e. the resulting information system, which is a human construction in which organisation and actors' culture will play a major part.

More generally, technology cannot be considered as the driver of the company, even if it has taken a strategic place as financial or human resources: "ERP systems promise to allow managers to retrieve relevant information from the system at any time and one knows that information is the key determinant of wealth in the modern economy. However, companies need to realise that if the ERP system is given too much control, then the foresight that is essential to adapt quickly to changing external factors can become blunt by an over-reliance on the technology driving the business. A lack of foresight will almost certainly mean loss of business" (Davenport, 1998).

A study of the scientific literature from 1998 to 2002 shows that the standardisation often intended in projects and the need to lead to a positive result do not ensure gaining competitive advantages from the ERP itself. On the contrary, it lies in the quality of the implementation, in the refinement of process definition, and in the alignment of the ERP system to the strategy of the company. As Beard and Sumner (2004) say: "An examination of the existing research suggests that ERP systems may not provide a competitive advantage based upon the premises of system value, distribution, and imitability. This is largely due to the "common systems" approach used for the implementation of most ERP systems. Instead, the source of competitive advantage may lie in the careful planning and successful management of ERP projects, refinement of the re-engineering of the organisation, and the post-implementation alignment of the ERP system with the organisation's strategic direction."

The benefit comes from the use of the ERP system and not from its implementation alone. Many authors agreed with this assessment: Donovan (2000) states that unarguably, ROI comes from process improvements ERP supports, not from new ERP software. Tomas (1999) emphasises that the true reason is not knowing if the firm has the best tool but wondering if it trains the best artisans to use it efficiently. "In essence, ERP deployment in itself saves nothing and does not improve anything. It is people and processes that create benefits" (Kumar et al., 2003). The use of ERP systems becomes as important as the system itself, as shown by Corniou (2002): "Did ERP systems deeply change a firm's life? No, it is not the tools that change but rather the human being, who learns to know and use them better. We must take a fresh look at uses and work context, and above all use grey matter that will remain the raw material of companies."

Only a quarter of firms describe the system appropriation as high, and more that a third as poor (Labruyere et al., 2002). A detailed research analyses the reason why a significant number of employees do not use the ERP system and bypass it (Calisir and Calisir, 2004). Usefulness factors are studied in order to measure the ERP contribution to user satisfaction. Perceived usefulness and ease of learning are decisive factors for user satisfaction. If the ergonomics of the tool itself is of course an important factor of use, the authors underline that it is not enough to develop its use. This presupposes that users understand the feasibility and the usefulness of the required effort. Therefore, the human factor is decisive in project achievement and usage effectiveness conditions. Many studies reinforce experiments and highlight that employee's confidence is variable, depending notably on the position in the project: decision-maker, project-team member, user, service provider. The construction of such a required confidence for end-users needs to consider different mechanisms or strategies like reputation, integrity, involvement, predictability, user concern, supervision sharing, availability… (Lander et al., 2004)

That is why it is essential for ERP project control to propose a framework able to characterise and evaluate existing uses and offer a usage improvement methodology. This is the purpose of the maturity model proposed in this chapter based among others on several experiment survey results.

10.3 Synthesis of ERP Surveys

Since 2000, numerous reviews on ERP projects have been undertaken in Europe or in USA. Some are quantitative or qualitative surveys, others are based on case studies. This section presents a synthesis of different surveys about management issues in ERP implementation projects. The objective is to bring out some relevant elements for the problem of optimisation of ERP use.

10.3.1 Investigations into ERP Projects

10.3.1.1 Survey Characteristics
The surveys of ERP implementation in manufacturing firms aimed to analyse the return on experiments on the ERP project, to identify critical success factors and to investigate future developments. They are concerned with penetration of ERP, motives, implementation processes, functionalities implemented, major obstacles and operational benefits.

The first survey (denoted S1) was carried out by Mabert et al. (2003) between August and October 1999. They study the impact of organisation size on penetration of ERP, motivation, implementation strategies, modules and functionalities implemented, and operational benefits from ERP projects, by investigating 193 manufacturing companies in the USA that had adopted an ERP.

In the second one (S2) Olhager and Selldin (2003) from November 2000 to January 2001 surveyed ERP implementations in 511 Swedish manufacturing firms, concerned with ERP system penetration, the pre-implementation process, implementation experience, ERP system configuration, benefits, and future directions (response rate of 32.7%).

The third survey (S3) was carried out by Canonne and Damret (2002) from June 2001 to February 2002 among 3000 French companies with more than 100 employees (response rate of the order of 5%). Of the responses, 54% of the companies have implemented or were in the process of implementing an ERP, 13% were planning to implement one within the next 18 months and 33% had no plans for an ERP system for the near future.

The fourth investigation (S4) was conducted by Deloitte & Touche (Labruyere et al., 2002) from November 2001 to May 2002 among 347 small and medium-sized companies (mainly situated in the south-east part of France) which already had an ERP (response rate of 16.4%).

The fifth one (S5) was carried out by the Pôle Productique Rhône-Alpes (PPRA, 2003), from January to April 2002 among 400 medium sized industrial companies which had implemented (65%) or were in the process of implementing an ERP in the Rhône-Alpes region of France (response rate of 11.3%).

In the last investigation (S6), Kumar et al. (2003) investigated critical management issues in ERP implementation projects in 2002 among 20 Canadian organisations; they studied selection criteria (ERP vendor, project manager, and implementation partners), constitution of project team, project planning, training, infrastructure development, ongoing project management, quality assurance and stabilisation of ERP.

10.3.1.2 Synthesis of Main Results
From these surveys, we identified:

- Motives to implement an ERP system
- ERP module or functionality implemented
- Implementation strategies and parameters
- ERP adoption measurement
- Benefits and obstacles identified from returns on experiment

Two kinds of motives can be distinguished: technical motives and organisational or business motives. The former is made up of "solve the Y2K problem" (from 35% in S3 and S5 to 88% in S4), "replace legacy systems" (from 45% in S3 and S5 to more than 80% in S1 and S2) and "simplification and standardisation of systems" (from 59% in S3 to around 80% in S1 and S2). All companies had been operating with a patchwork of legacy systems that were becoming harder to maintain and upgrade; and the competitive pressures on them required increasingly more responsive systems with real-time integrated information that the legacy systems could not provide easily. The latter kind of motive is linked to the overall improvement of the information system or to company willingness to have a system able to improve its performance: increase performance (S4: 67%), increase productivity (S4: 54%), restructure company organisation (from 32% for S1 to more than 50% according to S1, S2, S3 and S4), ease of upgrading systems (more than 40% for S1 and S3), improve interactions and communications with suppliers and customers (from 39% for S3, to 75% for S1), gain strategic advantage (from 19% for S3 to 63% for S2 and 79% for S1), Link to global activities (from 35% for S3 to more than 55% for S1 and S2), response to market evolution (S4: 21%).

All the surveys show that financials, material management, sales and distribution, and production planning are the most frequently implemented modules. To a lesser extent, we find human resources management (about 40%), quality management (about 45%), maintenance management (about 30%), and research and development management (about 20%). Other functionalities were expected but are still absent, such as customer relationship or customer service management, and business intelligence.

Despite customisation possibilities of ERP systems, S2 and S3 reveal that most of the projects involved developments, mainly on production planning, sales, logistics and material management modules. S3 indicates that nearly half the firms had to adjust the system on the main functionalities, which generated an additional cost, about 12.3% of the budget. This finding is also identified by S1: the degree of customisation varies significantly across size of company; larger companies customise more. S6 adds that one of the major challenges an adopting organisation faces while configuring an ERP system is that software does not fit all their requirements.

The strategy used for the implementation is one of the most important factors in assessing the impact of an ERP system on an organisation. Strategies can range from a single go-live date for all modules (Big-Bang) to single go-live date for a subset of modules (Mini Big-Bang) to phasing in by module and/or site. According to S4 and S5 surveys, which more concerned small and medium-sized firms, "Big Bang" is the most frequent; this ratio is inverted in S3 where the presence of large companies is more important. Both S1 and S2 confirmed this finding.

Generally, ERP implementation times are often underestimated, and are exceeded in about 50% of the cases. The real duration corresponds on average to 150% of duration foreseen with one or even two adjournments of the start-up date. S4 informs us about the causes of the delays: customisation problems (17%), reliability of the tests (16%), data migration (12%), specific developments not ended (13%), elimination of "bugs" (9%), training not ended (8%), organisation not ready at the time of "go-live" (8%). These findings tend to confirm that while the Big-Bang approach usually results in the shortest implementation time, it is also the riskiest approach because it can expose the entire stability of a company in case of any problems.

Few studies investigate user satisfaction. S3 highlights different rates of satisfaction according to modules: the users are rather satisfied (rate superior to 75%) by finance/accounting, purchasing, materials management and sales management modules. Although they are often the subject of specific developments, production planning and logistics/distribution modules present a rate of weaker satisfaction. S4 measures the appropriation of the system by the users: 26% of the respondents considered it high, 39 % satisfactory and 35% weak.

Main benefits are synthesised on Table 10.1. Most of the perceived improvements correspond to the expectations, which companies had, but not necessarily in the same measure: the improvement of business indicators (number of backorders, stock shortage, and customer service rate) is far from being reached, and the surveys do not allow us to deduce the reasons. Furthermore, the reduction of direct costs (or IT costs), one of the main objectives of the projects, is not quoted in the major results obtained.

In synthesis, ERP improved the global vision of the company and the collaborative work permitting master data harmonisation, considerable reduction of information redundancy, and work in real time.

The surveys S3, S4, S5 and S6 also allowed identification of problems encountered by companies during ERP implementation. They are mainly related:

- to the adaptation of the company to the "ERP model" or of the ERP to company-specific requirements (about 76%),
- to the resistance to change (membership of the users, conflicts and social problems),
- to the resources of the project team (user availability, deficiencies of the integration teams, underestimation of the resources),
- and to the problems of data exchanges between the ERP and the existing information system (redundancy of information, choice of the data and messages to be exchanged).

Table 10.1. Synthesis of main benefits

	S1	S2	S3	S4
Availability of information / Quickened information response time	80%	75%	55%	71%
Increased interaction across the enterprise, Integration of business operations/processes	80%	70%	37%	n.a.
Improved lead time	60%	60%	24%	74%
Improved inventory levels and purchasing	60%	52%	33%	74%
Improved interaction with customers	60%	56%	18%	36%
Improved interaction with supplier	60%	55%	11%	59%
Reduced direct operating costs	40%	55%	5%	42%

It seems that the culture "management by objective" was not extended to ERP projects.

10.3.2 Investigations into ERP Optimisation Strategies

Ross and Vitale (2000) compared the stages of an ERP implementation to the journey of a prisoner escaping from an island prison. They identified five stages: (1) ERP design/the approach, (2) ERP implementation/the dive, (3) ERP stabilisation/resurfacing, (4) continuous improvement/swimming, and (5) transformation. Until now, researchers have investigated the ERP implementation process up to the stabilisation stage in order to identify the stage characteristics, critical success factors of implementation and best project practices. Fewer authors have worked on the two latter stages, i.e. optimising the use of the information system for company development and performance.

Canonne and Damret (2002) investigate further projects; several developments are operational or planned such as finite capacity planning (38%), business warehouse (38%), e-business (29%), CRM (27%), SCM (24%). Labruyere et al. (2002) were interested in the evolutions planned by companies; after the development of new functionalities (23%), the optimisation of the use of their system (10%) arrives in second position, followed by the change of version (7%), the internationalisation of the company thanks to the ERP (7%), the development of decision-making and the participant implication (7%). These findings show the wide variety of situations that may trigger interest in "optimisation" of an ERP system in the meaning given in the introduction.

Nicolaou (2004) identifies factors of a high quality "post implementation review" to ensure ERP implementation effectiveness. He compares them to critical success factors of ERP implementation. Using insights from case studies, he conceptually defines the construct of such post-implementation review quality from antecedent conditions during the implementation process and from potential outcomes. In fact, the effectiveness is more a process than a metric, and the capability of an organisation to maintain the effectiveness of the ERP can be evaluate as a "maturity level". The Capability Maturity Model proposed by the Software Engineering Institute defines clearly this notion of maturity level (CMMI 2007). April et al. (2005) proposed a model for software maintenance and Niessink and Vliet (1998) for IT service. Niazi et al. (2005) proposed a comparison of critical success factors in software implementation and process approach of CMMI. This approach of model of maturity applied to "information system use" is a relevant research issue for ERP effectiveness.

10.4 Towards a Maturity Model for ERP "Good Use"

Botta-Genoulaz and Millet (2005) present the results of a project launched in the Rhône-Alpes region (France) in order to identify best practices of ERP "optimisation" in companies, and their application context. They propose a typology of these "post go-live" situations for small and medium-sized firms.

The study was carried out between January and March 2003 among 217 manufacturing companies in the Rhône-Alpes region that have an ERP "stabilised" for at least one year (response rate: 14%). The survey questionnaire asked for information on ERP implementation and current use in the company: the respondent's and the company's characteristics, the ERP project characteristics and their initial contribution (motives, timelines, budgets, functionalities, benefits, user satisfaction), organisational characteristics (during and after stabilisation), needs for improvement/evolution and "post go-live" diagnostic. It concerns mainly medium-sized companies (annual revenue between 15 M€ and 300M€, from 130 to 1,400 employees) that predominantly belong to a group (76.7%). These projects, introduced by the Head Office in 73% of the cases, were characterised by an average budget of 2,57M€, of which 1,37M€ of external services. The average implementation time was about 22 months, while it was estimated as 17 months: 63% of the firms underestimated this parameter. Big-Bang strategy was used in 74% of the cases.

More than 75% of companies consider themselves (very) satisfied with the project. For 52% of them, the ERP encompasses more than 75% of their entire information system; but most of them have one at least functionality covered by a specific development. Benefits measured agree with previous studies.

Ninety percent of the respondents consider it necessary to optimise the conditions of use and functioning of their ERP system. Motives for optimisation deal with:

- better use and exploitation of the ERP,
- expected results not reached,
- insufficient knowledge of the system installed,
- evolution of needs,
- evolution of the environment.

Companies are looking at evolutions such as deployment of new functions, optimisation of existing tools utilisation, upgrade of version, implementation of Business Intelligence solutions, and geographic deployment on multi-site companies. After implementation of the ERP, the organisation of the company was adapted by the creation of an ERP centre of competence, formalisation of owners of processes, and definition of the operational roles in the processes. This confirms that companies have organised themselves to support optimisation actions on their ERP based information system.

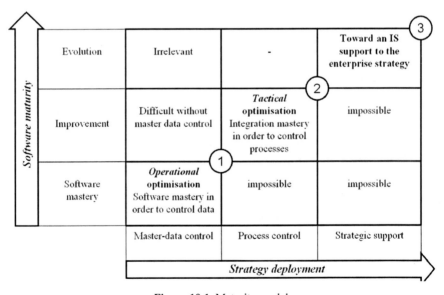

Figure 10.1. Maturity model

Regarding the motives listed above, cases 1 and 3 come close to the need to master the ERP system; one can talk about corrective optimisation required by the weakness of the initial ERP project. Cases 4 and 5 come close to objectives for internal improvements or resulting from external changes. Case 2 can come close to both, depending on the considered results.

10.4.1 Model Characteristics

A detailed analysis of identified traps, expected improvments and optimisation motives presented by Botta-Genoulaz and Millet (2005) leads to the identification of two axes to measure perfect command and control of the information system; each is split into three levels. The first axis entitled "Software maturity" relates to the good use of these systems from the point of view of their proper efficiency, and is separated into Software mastery, Improvement, and Evolution. The second, entitled "Strategy deployment", relates to the contribution of the information system to the performance of the company itself, to its global efficiency; it is separated into Master-data control, Process control, and Strategic support.

Table 10.2. Software maturity axis

Level	Alerts	Actions
Software Mastery	• Non-appropriation of the system by the users • Unsatisfactory operational execution • Insufficient speed/ability to react • Insufficient system response time • The users create parallel procedures • No documentation on parameters, data, data management procedures	• Additional training of the users • Create a competence centre • Empowerment of the users in their role and in their duty (user's charter, quality indicators) • Stabilisation of the execution (indicators with follow-up of objectives)
Improvement	• The full ERP potential is not used • Results not reached, expectations unsatisfied • The standard system installed does not fit all requirements • The number of office automation utilities increases • The procedures are too heavy	• Definition of performance indicators, business indicators • Improvement and automation of the reporting • Rethink the roles to simplify the procedures • Implement the functions that are not yet used
Evolution	• Context "multi-activities", international firm • Reorganisations, technological changes • Need for (analytics)reporting • Outside integration: B to B • Bar-code integration • Version upgrade • Software maturity depreciation	• Standardisation on several sites/activities • Version upgrade • Address the ERP / environment technological evolution (EAI) • Develop business intelligence systems • Implementation of enterprise architecture (application mapping)

Figure 10.1 illustrates the two-axes maturity model and the different possible stages depending on the two axes. This model proposes a synthetic vision of the process of optimisation. It underlines the constraint of coherence between both axes: the information system cannot support company strategy without being

mastered as a "tool". Certain situations are consequently impossible (control of the processes without mastery of the software).

Every level is defined by alert criteria allowing recognising, and by the typical actions of improvement to be implemented at this level. These alert criteria and improvement actions are presented in Table 10.2 for the software maturity axis and in Table 10.3 for the strategy deployment axis.

Table 10.3. Strategy deployment axis

Level	Alerts	Actions
Master-data control	• Numerous erroneous technical data • Messages of ERP not relevant (stock shortages, rescheduling in/out MRP, purchase proposals) • Numerous manual inventory corrections • Product lifecycle not improved, not integrated in the IS (revision)	• Cleaning of the migrated data • Define responsibility for data • Assert the uniqueness of the data in the whole company • Indicators of data control • Maintain a business project team with a plan of action to master data
Process control	• Conflicts between services on procedures • Contradictions between local and global indicators • Results not reached, needs unsatisfied • Demands for improvement, for roles redefinition by the users • Higher expectations of customers and top management • No return on investment calculated	• Revise management rules in the company • Verify the appropriateness of the tool to the organisation • Rethink the roles to simplify the procedures • Define responsibility for processes • Strengthen the transverse responsibilities (indicators, communication)
Strategy support	• Business objectives not reached • Higher expectations of customers and top management • Changes of markets, of customer expectations • International extension • Management expectation concerning follow-up consultancy	• Modelling and optimisation of the supply chain • External integration: B to B • Implementation of application mapping • Business Process Management (modelling, process performance measure) • IT associated to business strategies

10.4.2 Towards a Guideline for ERP Use Improvement

On this basis, we can propose a process of optimisation (in the sense of a better use of the information system) in three stages, which produce an information system contributing to the strategy of the company (situation numbered "3" on the model, Figure 10.1). These three optimisation stages allow us to characterise three situations (numbered 1, 2 and 3), which are defined below.

- Situation 1 is described as a result of an operational optimisation centred on the good use of what exists ("Master the tools to master the data"). To reach situation 1, the information system is considered as a tool for production and broadcasting of data.
- Situation 2 is described as a result of a tactical optimisation centred on the best integration of what exists to allow more effective use (improve ERP use for better control of the processes). To reach this situation 2, the information system is considered as a support for the control of company operational processes.
- Situation 3 is defined as the maximum use of the information system focused on a strategic optimisation leading to modification of the positioning of the existing ERP in the information system strategy. The information is then a real component in defining the strategy of the company.

This approach matches the last three stages defined by Ross and Vitale (2000): stabilisation, continuous improvement, and transformation. Activities observed for the stabilisation stage are typically operational optimisation as defined in situation 1 (cleaning up data and parameters, resolving bugs in the software, providing additional training). During the continuous improvement stage, firms focus on implementing adding functionality such as bar coding, EDI, sales automation, etc., generating significant operating benefits, which fit with situation 2. Finally, situation 3 corresponds to the transformation stage, which aims to gain increased agility, organisational visibility and customer responsiveness. The process of optimisation proposed agree with the taxonomy designed by Al-Mashari et al. (2003), which illustrates that ERP benefits are realised when a tight link is established between implementation approach and business process performance measures.

10.5 Organisational and Temporal Heterogeneousness of an Information System

10.5.1 The Organisational Heterogeneousness

In most big companies, the ERP are expanded gradually in many "roll out" projects after pilot projects having define a "core model". The situation in a company is thus mostly a mixed situation where certain sites or activities are integrated into the corporate information system builds on an ERP while others continue to use legacy systems or local packages.

Similarly, after the project, the ERP package is in a given situation in a more or less extensive information system. Difficulties with the project led to exclusion from the ERP scope some functions possibly critical for the company, either because of weak matching of the standard functions of the ERP, or to reduce the load and cost of the project. These functions "outside ERP", for example cash management or customer risk management, are then often covered by various solutions according to subsidiaries and countries.

The objective of the initial corporate projects had often to answer the expectations of controlling and even centralisation of certain functions. The requirements concern mainly the strategic functions such as finance, but also in certain cases, supply chain management, project engineering and management, etc. In contrast, certain functions were sometimes considered as local, strongly linked to the particular business of the site, notably when the group is "loosely integrated". We can then have a corporate ERP scope excluding major functions such as production, managed in every site or subsidiary by local legacy tools that can be another ERP solution. Furthermore, the maturity reached by an entity in the organisation, for a functional domain or even a particular process is the result of a particular history. This maturity cannot be homogeneous in a company, but depends on each entity of the organisation (Figure 10.2).

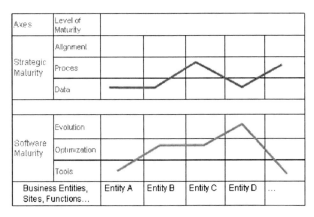

Figure 10.2. Heterogeneousness of the maturity in the organisation

This heterogeneousness is often strengthened by the history of the acquisitions, the merges or the transfers which characterise the large companies. This is the case of the AREVA group. Its nuclear part is historically managed with SAP, but some activities are managed with MOVEX and the "transport and distribution" part, arising from the repurchase of a division of the ALSTOM group is itself shared between SAP, BAAN, PRODSTAR. If the group posts an intention of rationalisation around SAP, it can be made only in a succession of projects which will have to justify each of their appropriateness, timeliness and, finally, return on investment. The maturity reached with an ERP on a process in a certain organisational context can be pushed aside by the arrival of a new ERP or the re-engineering of the same process considering a new economic or organisational context. The ABB group centralises its information systems in France around the ERP INFOR ERP LN (formerly BAAN) after having standardised it with SAP in other countries.

Finally, everything shows that this organisational heterogeneousness is not static. On the contrary, it is continuously transformed jointly in the life cycle of information systems, in the business cycles of the company and in implementation priorities of its commercial, industrial, financial, organisational, technological strategie. This heterogeneousness exists also at a lower scale in the mid-size companies, and even in small ones. Indeed, the constraints of budget and resources to drive the project are even stronger than in a big group. The compromises on project scope are thus even more necessary and lead to much differentiated situations. That can lead to a situation where a purchase service is operational with the ERP but unable to use completely the approval process of proposal orders because the supply function is not under the control of the ERP.

A model of maturity throughout the whole organisation can help to identify in a clear way the priorities of actions according to the functions or the sites, in a consistent way with the pursued global objectives and local capabilities.

10.5.2 The Temporal Heterogeneousness

The quality of the use of the ERP also has to take into account the "learning curve", and more generally the learning dynamics of the organisation, including the possible regression of maturity previously acquired. This can be the consequence of the loss of competence due to changes of staff in a weak or even non-existing knowledge management process. The "maturity" is never an acquired fact, never a static and structural characteristic of an organisation. It is necessarily changing with the company in its whole "life cycle", taking into account market trend and the positioning of the company, the technological cycles and their innovations breaks, the transformation of the logistical, and financial and commercial networks into which the company operates. Figure 10.3 shows such a maturity history.

During the stabilisation phase following the initial project, the maturity of the organisation grows through practical experience, but can decline after an extension of functional scope, which disrupts stabilised processes, or after deployment on a new site or entity. This "dynamics" of the organisation is a stake for the implementation phases itself. The failure of the ERP project at the DELL group was primarily due to the impossibility of building the project in the context of the growth rate and the strategic transformation of the company, which w as becoming the leader in online computer sales on the Internet (Trunick, 1999).

A main reorganisation of the company, especially after a merge or a transfer implies generally a redefinition of the information system strategy, with new decision-makers and a new context of commercial and management organisation. A frequent consequence is a loss of systems mastering, thus of maturity of the organisation. A main factor is the loss of competencies due to (sometimes numerous) loss of employees who were keys players in previous projects. Considering the cost of ERP projects, a strong financial and market position of the company and its capacity to finance such projects are obviously key success factors. However, the performance of a company can vary from one year to the next, and unplanned events can disrupt well-organised structures, so coming to disrupt the deployment of ERP systems in global projects planned over several years.

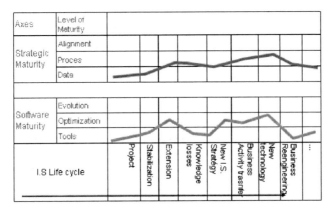

Figure 10.3. Maturity and life cycle of the information system

Finally, even without particular events, a well ground organisation is not static. The players evolve in their own careers, moving to new roles or companies. Even with a quality formalisation of the working procedures, player appropriateness must be maintained. That is the aim of a knowledge management process allowing an organisation to keep the operational processes under control. This appropriateness is attacked and often deteriorated by a loss of competences in the context of continuous evolution of the activity, which modify after time, the priorities of a process or make some particular cases critical when they had been considered negligible previously.

As any quality method, the maturity of use of an information system is not a continuously growing optimisation process, but requires a maturity management process for players who produce the information system in continuously changing companies. It has to lie within the geographical scope of the company (the services, the entities, the sites, the subsidiaries) and within the history (growth, reorganisation, merge, transfer). It requires evaluation tools to support dysfunction identification, and more generally audit actions, but also forward-looking tools allowing tracking of improvements, projects with scope and duration fitting with capacities and constraints of the company.

10.5.3 Dependences in the Model of Maturity

10.5.3.1 Integration and Coordination
The ERP projects answer needs of "informational" integration which are in fact the answer to the needs of coordination, of "organisational" integration. Any organisation can be characterised by a structure of hierarchical and functional links, which build the stability, the cohesion and the dynamics of the system. The reduction of the complexity by decomposition of a system in sub-systems leads generally to a hierarchical vision of the structure. This vision, however, has some problems:

- the hierarchical organisation of the decisions, decomposed in three levels, strategic, tactical and operational (Anthony, 1965);

- the rationality of the players, the nature of the information system, the autonomy of decision of the sub-systems (decentralisation of the decisions vs. control of the lower levels);
- the kind of integration (their contribution to a common purpose, their process of cooperation, coordination, etc.).

The decomposition of tasks constitutes only one of the foundations of the organisation. It leads one to identify the problem of the dependences between the various tasks. To minimise this problem, an approach of simplification of the coordination by changing the organisational structure is necessary (Thompson, 1967). It leads to the proposal of a mode of coordination adapted to increasing coordination difficulties:

- the coordination by rule inside the same structure allows regulating the so-called pooled interdependence;
- the coordination by planning of sequential activities (sequential interdependence);
- the coordination by mutual adjustment to answer the mutual interdependences (reciprocal interdependence).

This vision of simplification of the structure, to establish if possible coordination by rule will be completed by the work of Lawrence and Lorsch (1969). For these authors, there are two solutions to resolve these problems of coordination: reducing them by introduction of slack in the organisation, and increasing the capacities of integration of the organisation by the development of information systems. Their work enlightens the limits of a hierarchical control in a diversified environment. The adaptation to the context requires a decentralisation of the decisions associated with strong capacities of integration.

The search for a more effective coordination in the organisation thus leads to a stronger integration of information systems, vertically in the decision process, and horizontally in the geography of the organisation. This concept of integration is useful to describe the new modes of organisation based on narrower inter-individual, inter-functional and inter-companies relations. These relations are based themselves on a narrower coordination of the tasks, on cooperation and sharing of information, and finally on the decision-takings process. The management of the interdependences inside or outside a company leads to a complete informational integration (Geffroy-Maronnat et al., 2004).

The ERP, far from being only a marketing trend, corresponds to a deep transformation of organisations, which, by guaranteeing a functional interconnection, an inter-functional homogenisation and an adaptive opening, leads to "the old dream of a unique repository for the information system of the company" (Rowe, 1999).

According to the nature of organisations and the modes of control, the integration can take various forms, which can make the evolution and the change management difficult. The consistency between the organisational and informational dimension of integration is thus one of its factors of success.

10.5.3.2 Organisational and Informational Dependences

The notion of dependence is a key element of characterisation of the integration and a basis for measurement at the operational level of elementary entities (both of the organisation and the information system). Players are dependent when their tasks must be coordinated, either an a explicit way by a procedure linking the activities of some to the activities of others, or in an implicit way by the information system, which makes certain data or tools common.

Previous work allowed us to propose a general model of organisational and informational objects allowing characterization of the various forms of integration by leaning on the SCOR reference model (Stephens, 2001). This model uses various natures of objects to describe the supply chains and the characteristics of all elements of their performance: processes, functionalities, practices, information and metrics. Some objects, such as processes or metrics are clearly defined, codified, and classified into a hierarchy, others as functionalities and exchanged information are simply evoked in a descriptive way. We proposed a more general model (Millet, 2005) identifying the technical objects constituting the computer system, the informational objects constituting the information system, and the organisational objects. The UML model allows building an application facilitating navigation in all dependences through these different objects, and is presented in Figure 10.4.

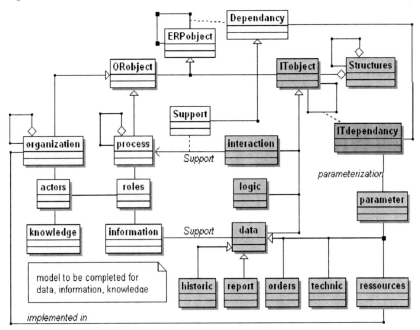

Figure 10.4. Class diagram of the informational and organisational objects

This model identifies the following objects and their dependences:

- organisational entities and the actors of these entities (organisation);
- roles defined for these actors (roles);

- processes which consistently linked the activities realised in the information system, the "transactions" of the package (process);
- allocation of the roles in the processes (relation roles–process);
- technical objects implemented in packages and software of the information system (IT object);
- data kept in the database of the information system (data);
- parameters setting the behaviour of the computer system (parameters).

The dependences between technical objects are identified from the software using cross-referencing tools. This can sometimes require more complex reverse-engineering tools.

The dependences between organisational and technical objects must be identified from a process model of the company, allowing one to analyse the matching between the information system and the processes and thus to the organisational entities which run and pilot these processes. These dependences are obviously less easily identifiable. Their formalisation requires work with the users and the manager of the organisation to model as clearly as possible the roles and the responsibilities. From this point of view, the dependences obtained will always be a more or less consensual "representation" of these dependences.

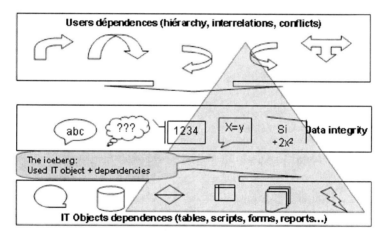

Figure 10.5. Dependences between informational and organisational objects

A less formal presentation of this model (Figure 10.5) classifies the dependences depending on objects in three levels, grouping together the organisational, informational, and technical objects. These dependences are then:

- organisational, which concern the collaborative practices, the hierarchical or functional relations between actors and entities;
- informational, which concern information systems, their data, their processes and the exchanges of information between these systems;
- technical, which concern the software and the technical systems supporting the flows of data and the metrics required for the control of organisations.

All these dependences can be represented in a graph, the "graph of dependences" which mixes the three organisational, informational and technical levels. The dependences "between levels" are critical because they model the contribution of a level as a "tool" to its level "of use". We can speak about adequacy of the technical infrastructure to the information system and about adequacy or alignment of the information system to the strategy and organisation of the company. The resultant graph can be clustered to obtain loosely coupled sub-graphs. Such a clustering, which can be treated by appropriate tools, allows one to identify sub-sets more or less correlated from the point of view of these dependences. Then, we can seek strongly integrated "domains" loosely coupled with the rest of the organisation. All the dependences between two domains represent the characteristics of the coupling between these domains. This coupling can be mainly informational in the case of a sales relation (for example based on orders and shipment). They can be technological in the case of a collaborative system sharing resources such as web services or interoperability components on an e-business platform. They can have a strong organisational content in the case of coordination of several entities in a group, or of collaborative practices (co-design, vendor managed inventory, supplier coaching, CPFR).

Such a representation of the various types of dependences allows studying their consistency to validate how a collaborative strategy at the organisational level is supported by a collaborative strategy at the informational and application level. It helps to build an improvement strategy identifying the risks and the necessary costs of work required by the intensity of the integration, measured with the number of dependences they imply. The model of maturity comes in this frame to allow estimating the capacity of the players to realise and to run this kind of integration.

10.6 Towards the Construction of a Learning Path

The model of maturity comes in three dimensions: use, organisational and temporal:

- the capabilities of using tools and the contribution of these tools to the performance, in other words, the global contribution of the computer system to the efficiency of the information system and the contribution of the information system to the business processes management and the performance of the company (dimension of "use");
- the scope of the organisation not only through its hierarchical structure but also through the more or less strong coupling that the organisational and informational dependences reveal ("organisational" dimension);
- the phasing of the evaluation and the action in the various life cycles of the company, taking into account its technological, commercial, financial transformations in the continuous market transformation ("temporal" dimension).

This model allows one to identify a "path of learning" defined by the scope of change management, an evaluation of the maturity of the entities in this scope through indicators and alerts, a plan of corrective tasks or improvements projects to reach a realistic maturity target (presented in Figure 10.6).

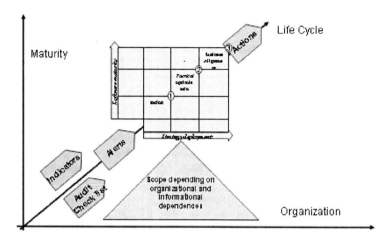

Figure 10.6. Path of learning in the organisation

The consistency and the aptness of the change scope in the global organisation has to be validated by the measure of its more or less coupling with the other parts of the organisation, by the identification of the external dependences which must be taken into account and processed in the change project, by the definition of indicators to measure the "endogenous" maturity of the players to minimise the disturbance which the external dependences can generate.

It assumes that the analysis of the dependences was realised for the whole information system and not only for the scope assumed for the project, to identify the "external" dependences with the project, which represent constraints.

The maturity is estimated on both axes "tools" and "strategy" to identify a realistic evaluation, through indicators concerning the operational, decision-making and strategic points of view.

The change project is defined from the level of maturity seen to reach a realistic target objective, taking into account the capabilities from an operational, decision-making and strategic point of view.

This model of maturity requires continuous definition of relevant indicators and corrective or enhancement tasks, to enrich a "repository" on the use of the integrated information system. This work cannot be realised in an academic way. It has to emerge from a global learning organisation, with the users of these integrated information systems. For this objective, we suggest pursuing the definition of this "model of maturity of organisations with integrated information system".

10.7 Conclusion

The stakes in the mastery of integrated enterprise systems are not limited to the phases of implementation or deployment. The "best use" of these information systems leads companies to new organisations and to continuous work on alignment of the strategy of the company. It is supposed to help in the evaluation of the role of the ERP system in the information system of the company to identify relevant improvement projects in a given situation.

From practices encountered in companies and from the results of various studies, we proposed a maturity model of the use of an ERP and a method of optimisation. This latter allows the identification of three levels: operational (the information system is considered as a tool for production and broadcasting of data), tactic (mastery of the operational processes and the integration between the functions) and finally strategic (in order to support the company in its transformations and evolutions). This model must be estimated in a more or less strongly integrated and heterogeneous organisation. It must be deployed on an "organisational" axis dependent on the scope and be used in a "life cycle" taking into account the transformations of the company in its commercial, financial or technological life cycles. The temporal axis represents the dynamics of the implementation of the model. In fact, it is the axis of change management.

The organisational and informational dependences which supports this integration must be identified and modelled to be able to propose a consistent scope of change project, loosely coupled with the rest of the company. These dependences can help in the definition of an optimisation path with which one can validate the feasibility. The construction of a repository of indicators, alerts, corrective tasks, and improvements projects, associated with a tool allowing the modelling of the dependences would supply a usable methodology of continuous improvement towards greater maturity of the alignment of information systems and business processes to the company strategy.

10.8 References

Abdinnour-Helm S, Lengnick-Hall ML, Lengnick-Hall CA, (2003) Pre-implementation attitudes and organizational readiness for implementing an Enterprise Resource Planning system. European Journal of Operational Research 146(2):258–273

Al-Mashari M, Al-Mudimigh A, Zairi M, (2003) Enterprise resource planning: A taxonomy of critical factors. European Journal of Operational Research, 146:352–364

Anthony RN, (1965) Planning and Control Systems: A Framework for Analysis. Division of Research, Graduate School of Business Administration, Harvard University.

April A, Huffman Hayes J, Dumnke R, (2005) Software Maintenance Maturity Model (SMmm): the software maintenance process model. Journal of Software Maintenance and Evolution: Research and Practice 17(3):197–223

Beard JW, Sumner M, (2004) Seeking strategic advantage in the post-net era: viewing ERP systems from the resource-based perspective. The Journal of Strategic Information Systems 13(2):129–150

Botta-Genoulaz V, (2005) Principes et méthodes pour l'intégration et l'optimisation du pilotage des systèmes de production et des chaînes logistiques. Mémoire d'Habilitation à

Diriger les Recherches, spécialité Productique et Informatique, INSA de Lyon & Université Claude Bernard – Lyon I

Botta-Genoulaz V, Millet PA, (2005) A classification for better use of ERP systems. Computers in Industry 56(6):572–586

Botta-Genoulaz V, Millet PA, Grabot B, (2005) A recent survey on the research literature on ERP systems. Computers in Industry 56(6):510–522

Bouillot C, (1999) Mise en place de Progiciels de Gestion Intégrée à l'occasion de fusions et cessions d'entreprises dans un contexte international. Systèmes d'Information et Management 4(4):3–20

Boutin P, (2001) Définition d'une méthodologie de mise en oeuvre et de prototypage d'un progiciel de gestion d'entreprise (ERP). PhD thesis. Ecole Nationale Superieure des Mines de Saint-Etienne

Calisir F, Calisir F, (2004) The relation of interface usability characteristics, perceived usefulness, and perceived ease of use to end-user satisfaction with enterprise resource planning (ERP) systems. Computers in Human Behavior 20(4):505–515

Canonne R, Damret JL, (2002) Résultats d'une enquête sur l'implantation et l'utilisation des ERP en France. Revue Française de Gestion Industrielle 21:29–36

CMMI, Architecture Team (2007) Introduction to the Architecture of the CMMI Framework. http://www.sei.cmu.edu/publications/documents/07.reports/07tn009.html September 13.

Corniou JP, (2002) Conclusion de la rencontre "Propos raisonnables sur les ERP". Revue Française de Gestion Industrielle 21(4):117–118

Davenport TH, (1998) Putting the enterprise into the enterprise system. Harvard Business Review 121–131

Deixonne JL, (2001) Piloter un projet ERP. Dunod, Paris

Donovan, RM, (2000) Why the Controversy over ROI from ERP? MidRange ERP, January 1999. http://www.midrangeERP.com (October 9).

Geffroy-Maronnat B, El Amrani R, Rowe F, (2004) Intégration du système d'information et transversalité. Comparaison des approches des PME et des grandes entreprises. Sciences De La Société 61:71–89

Gilbert P, (2001) Systèmes de gestion intégrés et changement organisationnel. Revue de Gestion des Ressources Humaines 41:21–31

Holland CR, Light B, (1999) A critical success factors model for ERP implementation. Software, IEEE 16(3):30–36

Krumbholz M, Maiden N, (2001) The implementation of enterprise resource planning packages in different organisational and national cultures. Information Systems 26(3):185–204

Kumar V, Maheshwari B, Kumar U, (2003) An investigation of critical management issues in ERP implementation: empirical evidence from Canadian organizations. Technovation 23:793–807

Labruyere E, Sebben P, Versini M, (2002) L'ERP a-t-il tenu ses promesses ? Deloitte & Touche, June

Lander MC, Purvis RL, McCray GE, Leigh W, (2004) Trust-building mechanisms utilized in outsourced IS development projects: a case study. Information & Management 41(4):509–528

Lawrence PR, Lorsch JW, (1969) Developing Organizations: Diagnosis and Action. Addison-Wesley Publishing Company, Reading, Mass. 01867

Mabert VA, Soni A, Venkataramanan MA, (2003) The impact of organization size on enterprise resource planning (ERP) implementations in the US manufacturing sector. Omega 31:235–246

Mabert VA, Soni A, Venkataramanan MA, (2001) Enterprise resource planning: common myths versus evolving reality. Business Horizons 44(3):69–76

Markus ML, Tanis C, (2000) The Enterprise Systems Experience – From Adoption to Success. R.W. Smud, M.F. Price (Eds). In Framing the Domains of IT Research: Glimpsing the Future Through the Past, 173–207

Millet PA (2005) A reverse Engineering Approach of Integration with ERP. Conference IFIP 5.7 Advanced in Production Management Systems (APMS), September 18–21, Rockville, MD, USA

Niazi M, Wilson D, Zowghi D, (2005) A maturity model for the implementation of software process improvement: an empirical study. Journal of Systems and Software 74(2):155–172

Nicolaou A, (2004) Quality of postimplementation review for enterprise resource planning systems. International Journal of Accounting Information Systems 5(1):25–49

Niessink F, van Vliet H, (1998) Towards mature IT services. Software Process: Improvement and Practice 4(2):55–71

O'Donnell E, David JS, (2000) How information systems influence user decisions: a research framework and literature review. International Journal of Accounting Information Systems 1(3):178–203

Olhager J, Selldin E, (2003) Enterprise resource planning survey of Swedish manufacturing firms. European Journal of Operational Research 146:365–373

Poston R, Grabski S, (2001) Financial impacts of enterprise resource planning implementations. International Journal of Accounting Information Systems 2:271–294

PPRA (2003) Enquête ERP. Pôle Productique Rhône-Alpes, www.productique.org

Ross JW, Vitale MR, (2000) The ERP Revolution: Surviving vs. Thriving. Information Systems Frontiers 2(2):233–241

Rowe F, (1999) Cohérence, intégration informationnelle et changement: esquisse d'un programme de recherche à partir des Progiciels Intégrés de Gestion. Systèmes d'Information et Management 4(4):3–20

Saint-Léger G, Neubert G, Pichot L, (2002) Projets ERP: Incidence des spécificités des entreprises sur les Facteurs Clés de Succès. Proceedings of AIM 2002, Hammamet, Tunisia, May 30–June 1

Simon HA, (1996). The Sciences of the Artificial (3rd edn.). MIT Press

Somers TM, Nelson KG, (2004) A taxonomy of players and activities across the ERP project life cycle. Information & Management 41:257–278

Stephens S, (2001) Supply Chain Operations Reference Model Version 5.0: A New Tool to Improve Supply Chain Efficiency and Achieve Best Practice. Information Systems Frontiers 3(4):471–476

Thompson JD, (1967) Organizations in Action. McGraw-Hill College

Tomas JL, (1999) ERP et progiciels intégrés. Dunod, Paris

Trunick PA, (1999) ERP: Promise or pipe dream? Transportation and Distribution. 40(1):23–26

11

A Cross-cultural Analysis of ERP Implementation by US and Greek Companies

Jaideep Motwani[1], Asli Yagmur Akbulut[1], Maria Argyropoulou[2]
[1]Grand Valley State University, Seidman College of Business, Department of Management
[2]Athens University of Economics and Business

11.1 Introduction

Enterprise Resource Planning (ERP) systems play an important role in integrating information and processes across departmental boundaries (Reimers, 2003; Klaus et al., 2000; Sankar et al., 2005). Organisations, especially in developing countries, have adopted these information systems extensively to overcome the limitations of fragmented and incompatible stand-alone and legacy systems (Huang and Palvia, 2001; Sharma et al., 2002; Robey et al., 2002). Even though the inherent appeal of ERP systems has not gone unnoticed in developing countries (Xue et al., 2005), ERP is still in its early stages in countries in Asia/Pacific, Latin America and Eastern Europe (Huang and Palvia, 2001; Rajapakse and Seddon, 2005) .

ERP systems are built on the best practices in industry, which represent the most cost-effective and efficient ways of performing business processes (Markus and Tannis, 2000; Sumner, 2004). The transfer of information systems like ERP, typically developed in industrialized countries, to developing countries is often marred by problems of mismatch with local cultural, economic and regulatory requirements. Considering that most ERP systems are designed by Western IT professionals, the structure and processes embedded within these systems reflect Western culture. Fundamental misalignments are likely to exist between foreign ERP systems and the natural and organisational cultures of companies in developing countries (Soh et al., 2000; Molla and Loukis, 2005; Rajapakse and Seddon, 2005). Yet, very little academic research has been conducted to investigate the influence of natural culture on ERP implementations (O'Kane and Roeber,

2004). Therefore, there is a need for research to examine generic and unique factors that affect ERP implementation success in culturally different contexts.

In this study, we will consider the social, cultural and contextual factors contributing to ERP success in USA and Greece by analysing a case example of ERP implementation in each country. These two countries differ significantly from each other based on Hofstede's classification of national culture (Hofstede, 1991, 2001). Hofstede's classification of national culture has identified four dimensions of culture: power distance, uncertainty avoidance, masculinity/femininity, and individualism/collectivism. The cultural differences in the USA and Greece along these four dimensions can significantly impact the success of the ERP implementations. Therefore, we believe that by investigating ERP implementations in these two different countries, we will deepen the understanding of ERP implementations and provide suggestions as to how managers can increase the rate of success of ERP implementation in culturally different contexts. In this study, we use Hofstede's cultural model because it has proven to be stable and useful in numerous studies across many disciplines.

11.2 Literature Review

The literature section comprises three parts. The first part provides a summary of the ERP literature in general and identifies the critical factors for success. In the second part, specific cultural studies related to ERP are examined. In part three, the cultural dimensions used for examining the differences in ERP implementation between US and Greek companies are elaborated on and four propositions that we plan to investigate are developed. .

11.2.1 Prescriptive Literature on ERP

Publications on ERP systems have focused on many different research issues. After an extensive review of the literature, Esteves and Pastor (2001) classified ERP system research into the following categories: general ERP research (overview of ERP systems, research agendas; motivations and expectations; and proposals on how to analyse the value of ERP systems), adoption, acquisition, implementation, usage, evaluation, and education. Within the implementation category, several studies have been conducted to examine the factors that facilitate or inhibit the success of ERP implementation projects. For example, Brown and Vessey (1999) identified ERP implementation variables that may be critical to successful implementation through literature review and incorporated those variables into a preliminary contingency framework. Parr and Shanks (2000) built a phased project model consisting of planning, set-up, and enhancement phases and then identified the critical success factors that are important within each phase. Esteves and Pastor (2000) created a unified critical success factors model for ERP implementation projects. Murray and Coffin (2001) identified frequently cited ERP Critical Success Factors and compared the identified factors with actual practice using two case studies. Allen et al. (2002) identified ERP critical success factors for public organisations. Al-Mashari et al. (2003) developed a taxonomy of ERP critical success factors to demonstrate the linkages between ERP critical success factors, ERP success and ERP benefits. Umble et al. (2003) identified success factors, software selection steps, and implementation procedures critical to a successful ERP implementation. Tatsiopoulos et al. (2003) proposed a structured risk management approach for successful implementation of ERP systems, and examined its application.

Somers and Nelson (2004) identified and tested the relative importance of the key players and activities across the ERP project life cycle, which affect the success of these projects. Motwani et al. (2005) identified the factors that facilitated the success of ERP implementations. The authors also examined the factors that initially inhibited the success of the implementation process and explained how these barriers were overcome. Gargeya and Brady (2005) content analysed secondary data pertaining to ERP implementations to identify the facilitators and inhibitors of implementation success. Tsai et al. (2005) identified critical failures factors in ERP implementations and provided suggestions as to what to focus on to increase the rate of success.

Table 11.1 (at the end of the chapter) summarizes the major recent studies that focus on critical success factors for ERP implementations. The methodological approach of each study as well as the critical success factors identified in each study are provided.

11.2.2 Cultural Studies on ERP

Most of the existing studies that investigate the success factors for ERP implementations focus on projects that have been carried out in North America and Western Europe (Davison, 2002). These studies contribute greatly to our knowledgebase; however, one major limitation of the relevant literature has been the lack of studies that focus on implementation issues in developing countries. As such, more recently, recognizing the fact that national culture can impact the adoption and successful implementation of western based ERP software, researchers have started to examine the ERP implementations in other countries, particularly in Asia.

For example, Soh et al. (2000) discussed the cultural misfits of ERP packages from a Singaporean perspective. Huang and Palvia (2001) identified a range of issues concerning ERP implementation by making a comparison of advanced and developing countries. Davison (2002) compared educational ERP system implementation practices in North America and Hong Kong. Reimers (2003) investigated the crucial implementation process and context variables which warrant closer study of ERP enabled organisational change in China. Liang et al. (2004) investigated the five companies that attempted to implement foreign ERP systems with unsuccessful results and identified several problems that resulted in failure. Martinsons (2004) investigated the ERP implementations in China and concluded that there is a poor fit between ERP systems and traditional Chinese management systems. O'Kane and Roeber (2004) focused on an ERP implementation in a Korean company and determined what impact natural culture has on the implementation process of ERP systems.

Utilizing Hofstede's dimensions of national culture, Rajapakse and Seddon (2005) investigated six ERP implementations in Sri Lanka. The findings revealed a clash of cultural forces between the culture embedded in Western products and the culture of Asian ERP adopters.

Table 11.2 (see end of chapter) summarizes the major studies that focus on the role of culture in ERP implementations. The methodological approach of each study as well as the findings are provided.

11.2.3 Hofstede's Cultural Dimensions and Propositions

Hofstede defines organisational culture as *"the collective programming of the mind, which characterize the members of one organisation from others,"* (1991, p. 237) and national culture as *"the collective programming of the mind which distinguishes the members in one human group from another"* (1991, p. 21). Based on an extensive study of national cultures across more than 70 countries, Hofstede (2001) developed a model that identifies the following four primary dimensions to assist in differentiating cultures: Power Distance, Uncertainty Avoidance, Masculinity and Individualism. These dimensions are discussed below:

Power Distance (PD): This dimension focuses on people's beliefs about unequal distributions of power and status, and their acceptance of this inequality. In countries that have a high power distance culture, individuals with positions/title inherit considerable power and employees in these cultures tend to accept centralized power and depend heavily on their superiors for direction since they are less likely to be involved in any decision making. On the other hand, in countries that have a low power distance culture, individuals expect to be involved in decision-making and are less likely to accept centralized power. In other words, employee participation is part of lower power distance culture.

Uncertainty Avoidance (UA): Hofstede defines this second dimension as the "extent to which the members of a culture feel threatened by uncertain or unknown situations" (Hofstede, 1991, p. 113). In high Uncertainty Avoidance cultures, organisations and individuals are so used to doing things in a traditional way that they tend to resist new technology because of the potential risk associated with it. On the other hand, in low Uncertainty Avoidance cultures, there is less need for predictability and rule-dependency, and therefore, these cultures are more trusting than their counterparts (De Mooij, 2000) and are willing to adopt and implement new technologies in their working tasks (Maitland and Bauer, 2001; Veiga et al., 2001).

Masculinity/Femininity (MAS): According to Hofstede (1991), "masculinity" pertains to societies where social gender roles are clearly distinct (i.e., "masculine" countries value assertiveness and focus on material success, while "feminine" countries value modesty, tenderness, and quality of life). Also, the quality of life in feminine cultures is extended to workplace as well (De Mooij, 2000). This is not so true in masculine cultures, where a stricter task orientation prevails.

Individualism/Collectivism (IDV): Under this particular dimension countries are either labelled "individualistic" or "collectivistic." According to Hofstede (1991, p. 114), "Individualism pertains to societies where individual ties are loose and everyone is expected to look out for themselves and their family. In collectivist societies, on the other hand, people are integrated at birth into strongly cohesive in-groups, and group loyalty lasts a lifetime." In other words, collectivist societies are integrated and individuals from these societies think in "we" terms but in individualist societies, individuals think in "me" terms.

Based on the above descriptions of Hofstede's four dimensions, we expect to find both similarities and differences in the ERP implementation process between the US and Greek case study companies.

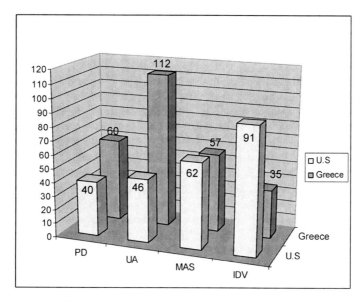

Figure 11.1. Comparison between USA and Greece

Figure 11.1 compares USA and Greece along the four dimensions of national culture identified by Hofstede. More specifically, we offer the following propositions:

Power Distance (PD): Considering that there is a 20 point spread between USA (40) and Greece (60) we expect to see differences in each company's approach to ERP implementation. Since USA is lower in PDI we expect to find examples of close working relationships between management and subordinates in their ERP implementation as compared to Greece.

Uncertainty Avoidance (UA): We expect to find clear differences in the UAI dimension, given the huge contrast between the USA (46) and Greece (112) scores. Based on the scores, we expect that people in Greece will be reluctant to make decisions and would prefer a very structured work routine compared with those in the USA.

Masculinity/Femininity (MAS): Since both USA (62) and Greece (57) are within 5 points of each other, we expect them to behave very similarly. Since both countries are above the world average of 50, we can expect that a higher degree of gender differentiation of roles exists and that the male dominates a significant portion of the society and power structure.

Individualism/Collectivism (IDV): We expect to find clear differences in the IDV dimension, given the stark difference between the U.S. (91) and Greece (35) scores. Based on the scores, we expect that there will be a greater level of tolerance for a variety of ideas, thoughts, and beliefs in the US company.

11.3 Methodology

A case study methodology was utilized in this study. According to Yin (2003), "A case study is an empirical inquiry that investigates a contemporary phenomenon within a real-life context, specifically when the boundaries between phenomenon and context are not clearly evident." Since our study attempts to investigate a contemporary phenomenon within a real-life context, i.e. the ERP adoption process (specifically, we focus on the cultural issues critical to ERP success at the case study company), we decided to utilize the case study methodology.

Data were collected primarily through interviews, observations, and document analysis. When available documents related to each organisation and the implementation project, such as mission statements, feasibility studies, reports, meeting minutes, RFPs, project plans, etc., were reviewed. Interviews were conducted with key players in the ERP implementation projects including members of the top management, functional area representatives, information technology (IT) professionals and end-users.

Case data were analysed to determine the factors influencing ERP implementation. The researchers individually and collectively analysed the data to allow the case to be viewed from different perspectives (Dubé and Paré, 2003; Eisenhardt, 1989).

11.4 Case Analysis: Implementation and Discussion

This section comprises three parts. In parts 1 and 2, a brief introduction to the ERP implementation at each case study company is provided. In part 3, Hofstede's four cultural dimensions are used for analysing and comparing the implementation processes used by the two case study companies.

11.4.1 US Case Study Company: US Global Energy Corporation

US Global Energy Corporation (a pseudonym for the actual company name) is a large global energy company with revenues exceeding 50 billion dollars. The company is engaged in exploration, production, refining, marketing, and distribution of energy products and technologies.

The management of US Global Energy Corporation recognised the need for an integrated system to manage the increasing complexities of its business. Prior to the implementation of ERP, the company used a number of separate systems to manage the enterprise but found that the lack of integration and increased complexity caused by growth were rendering these systems inadequate. Top management decided on a revolutionary "all-at-once" replacement of selected legacy systems with an ERP system. While the ERP system did not replace all of the legacy systems (a conscious choice made by the company), it did greatly simplify the processes and flows of information throughout the company.

The company chose to implement SAP's R/2 solution in its Chemicals division and to pilot the software in its chemical factory. Consultants from SAP worked closely with their SAP Competence Centre and US Global Energy Corporation's IT department in a series of sizing exercises to determine the appropriate equipment, storage, availability and backup needed under various scenarios.

11.4.2 Greek Case Study Company: Greek Coating Corporation

Greek Coating Corporation (a pseudonym for the actual company name) is one of the most important manufacturers of industrial coatings of high quality standards in the Greek market. The company's unique competence lies in the President's leadership style and motivated staff. Today, they are considered a reliable and dynamic enterprise with constant growth, based on high technical specialisation, strong customer relations, and solid technical support. The company employs approximately 100 people in production, offices and sales. Two production lines are used for "the make to stock" products having well known demand and one line is dedicated to the make to order production of specific products which concern unique colours for special customers with particular technical specifications. The products are distributed through the company's own logistics network.

In recent years, the company has expanded its international activities by means of its subsidiaries and network of sales. However, the rapid growth found their production system unable to fulfil demand and as a consequence top management decided to proceed with an extra production line, which did not help significantly as the real problem lay in inadequate overall master planning. For that reason the President decided to proceed with the implementation of a packaged ERP software in their attempt to have better planning, reduced lead times and faster information to customers. He purchased the same system from SAP as his main competitor, operating in Italy. However, during implementation, the problems discovered were so numerous that they had to either re-engineer their process or proceed to major customisations, which caused anxiety, denial and a tremendous increase in costs.

11.4.3 Discussion

In this section, we compare our findings for the US and Greek companies with Hofstede's four cultural dimensions.

Power Distance (PD): According to Hofstede (1991), moderately low power - distance countries, such as the USA, show limited acceptance of power inequality and less dependence of subordinates on bosses. In such countries, we not only expect to find examples of close working relationships between management and subordinates but also examples of assertive behaviour by subordinates, such as defining their own work tasks. Our findings for this dimension indicate high concurrence with Hofstede's general description of moderately low PD for USA and higher PD score for Greece. For example, a key success to ERP implementation at the US company can be attributed to the formation of cross-functional teams by top management. Three crucial teams were assembled to ensure successful implementation – a strategic thinking team, a business analysts group, and an operations group. On the other hand, the Greek company's initiative for the ERP system came directly from the President's rushed decision. The President believed so much in this change that he tried to persuade all his employees of the necessity for rapid ERP adoption. He asked his managers to become the change agents and to directly report to him. Since there was pressure to complete the implementation in a short period of time, no formal teams were created. The managers and employees didn't question the President and just followed the directions that were issued.

Uncertainty Avoidance (UA): According to Hofstede (2001), in high Uncertainty Avoidance cultures such as Greece, organisations and individuals tend to resist to technological change because of the potential risk associated with it. They feel more comfortable in doing things in a structured manner. On the other hand, in low Uncertainty Avoidance cultures, such as the USA, individuals are more willing to adopt and implement new technologies. Our findings for this dimension indicate partial concurrence with Hofstede's general description of low UA scores for USA and very high UA scores for Greece. For example, the US company was very successful in its ability to take all employees in their fold. Employees were willing to allocate a large amount of their time to the project. They were aided by training sessions that were available both day and night. The open communication encouraged by management gave users a sense of ownership of the system. Also, the teams worked very closely with the ERP vendor during the implementation process. They even allowed vendor consultants remote access to their system. When any problems were discovered, managers would meet with their vendors to discuss the same and contact vendor consultants for fixes.

On the other hand, at the Greek company, the President enjoyed the employee's commitment as he had always helped everyone advance their career and paid them well with high salaries and productivity bonuses. The whole company, unlike other Greek companies in general, was always more flexible and vigilant for new ideas, policies and changes. Therefore, when the President decided to implement an ERP system, there was really no resentment from the managers and employees. They trusted the President and showed support for his initiative even though it involved a drastic change to the way they did their work. Also, one would normally expect, more planning and attention to detail in the Greek company. However, this did not occur. They unfortunately underestimated the complexity and pitfalls of the ERP project and struggled through the implementation process.

Masculinity/Femininity (MAS): According to Hofstede (1991), strong "masculine" countries value traits like authority, assertiveness, performance and material success. "Feminine" countries, on the other hand, value modesty, tenderness, and quality of life. Since both USA (62) and Greece (57) are "masculine" cultures and are within 5 points of each other, we expected them to behave similarly. Like a true "masculine" society, the US company developed and followed an outcome and process-oriented approach to ERP implementation. They devised a strategic plan tied in with its ERP and business process change efforts that focused on incremental improvements. For example, the project leader in the strategic thinking team was tasked with developing the master plan and implementation deadline. The strategic thinking team determined that the finance function (Configurable Enterprise Financials including sub-modules for accounts payable, accounts receivable, general ledger and fixed assets) would be the first to be converted to the new system, giving users time to get used to the new system. Converting the operations function to the ERP system would follow. The modules were selected in conjunction with the determination of which facilities would be implemented first. On the other hand, the President of the Greek Company used his authority and assertiveness to initiate the ERP implementation process. The strategic decision-making depended mainly on the President's own critical thinking and experience. The ERP system selection process was based on the President's decision to do better than his competitor. Also, there was no resistance and denial from his managers and staff as they all trusted his insight and risk taking policies, which until then had proved beneficial.

Individualism/Collectivism (IDV): The intent of this scale is to measure whether the people prefer to work alone or in groups. Under this particular dimension countries are either labelled "individualistic" or "collectivistic." As mentioned earlier, collectivist societies are integrated and individuals from these societies think in "we" terms but in individualist societies, individuals think in "me" terms. Since USA measures lower on this scale, we expected there to be a stark difference between the USA and Greece in this dimension. However, this was only partially true. Our findings in this dimension demonstrate both concurrence with and differences from Hofstede's conclusion. Our findings show characteristics of both individualism and collectivism in both countries. For example, in individualistic cultures, like the USA, ERP is viewed as useful because it enhances the performance of the individual in spite of being viewed as a collaborative system. A comment by a Greek respondent illustrates the individualist nature as well: "I was not trained in ERP and everything I have learned I have taught myself." While the dominant characteristic seen here is individualism, both cultures also displayed collectivist values. For example, both company executives and employees described the value of shared information provided by ERP. While we expected to find greater differences in this dimension, based on Hofstede, the findings are interesting with respect to the similar comments made by the interviewees representing both cultures.

11.5 Conclusions

This study compares US and Greek cultures with regard to the implementation of ERP systems, which according to our knowledge has not been investigated before. Hofstede's cultural theory suggests that US culture is quite different from Greek culture in at least three of the four dimensions. Overall, our findings are consistent with Hofstede's in most of the dimensions.

Based on the results of our comparative cross-cultural case analysis, we can conclude that, in spite of the cultural differences, there exist some common underlying threads that are critical for ERP success. These threads or critical factors are consistent with the findings of prior research studies and are not culturally bound. First, according to Lee (2000), top management needs to publicly and explicitly identify the ERP project as a top priority. In both cases, this was true. However, in the US company, the strategy was well-planned and implemented a well-planned strategy. As such, they were more successful as the top management was able to develop a shared vision of the organisation and to communicate the importance of the new system and structures more effectively to their employees. Second, a clear business plan and vision to steer the direction of the project is needed throughout the ERP life cycle (Amin et al., 1999). The US company had a clear business model of how the organisation should operate behind the implementation effort. On the other hand, the Greek company did have a plan; however, since the plan was President driven, they ran into several obstacles. Third, a project champion is critical to drive consensus and to oversee the entire life cycle of implementation (Bingi et al., 1999). In the US company, a high level executive sponsor was selected as the project leader while in the Greek company, the President served as executive sponsor and project leader. Fourth, according to Holland and Light (1999), organisations implementing ERP systems should work well with vendors and consultants on software development, testing, and troubleshooting. In the US case study, the project teams worked very closely with vendors to obtain inter-organisational linkages, while in the case of the Greek company, the consultant worked closely with the President. Lastly, the progress of the ERP project should be monitored actively through set milestones and targets. According to the experts interviewed, process metrics and project management tools and techniques were used to measure progress against completion dates, costs, and quality targets in the US company but were used minimally by the Greek company.

The overview of culture and the cultural framework that is provided in this paper clearly illustrates the importance of culture, and the impact that each of Hofstede's dimensions has on ERP implementation. In conclusion, we would like to concur with Xue et al. (2005) that "While we recommend ERP vendors and implementing companies to pay attention to the cultural and non-cultural factors we identified to increase the likelihood of achieving ERP success, we would like it to be recognised that addressing these factors at the beginning of an ERP project cannot guarantee later success." This is especially true when ERP is implemented in different cultures. ERP implementation is a dynamic process and therefore, problems can arise at any phase of the process. However, to enhance the success rate, we strongly believe that a cautious, evolutionary, implementation process backed with careful change management, network relationships, and cultural readiness must be utilized.

Table 11.1. Major studies examining critical success factors for ERP implementations

Study	Methodology	Critical Success Factors Identified
Brown and Vessey (1999)	Case Study (2 organisations – preliminary results)	*Identified ERP implementation variables that may be critical to successful implementation through literature review and incorporated those variables into a preliminary contingency framework:* • Top management support • Composition and leadership of the project team • Attention to change management • Usage of 3rd party consultants • Management of complexity by: extent of process innovation, degree of package customisation, conversion strategy
Holland and Light (1999)	Case study (8 organisations)	*Identified ERP critical success factors:* • Strategic: legacy systems, business vision, ERP strategy, top management support, project schedule/plans • Tactical: Client consultation, personnel, business process change and software configuration, client acceptance, monitoring and feedback, communication , troubleshooting
Esteves and Pastor (2000)	Literature Review	*Created a unified critical success factors model:* • Organisational/Strategic: sustained management support, effective organisational change management, adequate project team composition, good project scope management, comprehensive business re-engineering, adequate project sponsor role, adequate project manager role, trust between partners, user involvement and participation • Organisational/Tactical: dedicated staff and consultants, appropriate usage of consultants, empowered decision makers, adequate training program, Strong communication inwards and outwards, formalised project plan/schedule, reduce troubleshooting

Table 11.1. (continued)

Study	Methodology	Critical Success Factors Identified
Esteves and Pastor (2000) (continued)		• Technological/Strategic: avoid customisation, adequate ERP implementation strategy, adequate ERP version • Technological/Tactical: adequate infrastructure and interfaces, adequate legacy systems and knowledge

Parr and Shanks (2000)	Case study
	
	(2 organisations)
		Recommend a phased model approach to ERP implementation projects and investigated which critical success factors are necessary within each phase of this model:
		• Planning phase: management support, champion, commitment to change, vanilla ERP, best people full-time, deliverable dates, definition of scope and goals.
		• Project phase:
		  ○ Set-up: Management support, balanced team, definition of scope and goals, champion, vanilla ERP, deliverable dates, definition of scope and goals
		  ○ Re-engineering: balanced team, definition of scope and goals, empowered decision makers, management support,
		  ○ Design: best people full time, vanilla ERP, management support, commitment to change, deliverable dates
		  ○ Configuration and testing: best people full time, vanilla ERP, management support, balanced team
		  ○ Installation: management support, commitment to change, balanced team, best people full time
		• Enhancement phase: not identified

Murray and Coffin (2001)	Case study (2 organisations – 1 private sector, 1 government organisation)	*Identified frequently cited ERP critical success factors and compared the identified factors with actual practice using two case studies:* • Executive support is pervasive and accountability measures for success are applied • Business process/rules are well understood and functional requirements built from these processes are clearly defined before selecting an ERP product • Minimal customisation is utilised • ERP is treated as a program, not project • Organisation wide education and adequate training are provided • Realistic expectations in regards to ROI and reduced IT/IS costs exist • Realistic deadlines for implementation are set

Table 11.1. (continued)

Study	Methodology	Critical Success Factors Identified
Roseman et al. (2001)	Literature Review	*Developed a priori model for process modelling success factors derived from the literature:* • Modelling methodology, modelling language, modelling tool, modeller's expertise, modelling team orientation • Project management, user participation, top management support
Allen et al. (2002)	Case study (4 higher education institutions)	*Identified ERP critical success factors for public organisations:* • *Strategic: project schedule/plans, ERP strategy, mission, top management support* • Contextual: organisational culture, constructions of past, technological implementations, political structures • Tactical: relationship and knowledge management, business process changes and software configuration, technical tasks, client acceptance, monitoring and feedback, troubleshooting, • Communication
Al-Mashari et al. (2003)	Literature Review	*Developed a taxonomy of ERP critical success factors to demonstrate the linkages between ERP critical success factors, ERP success and ERP benefits:* • Setting-up: management and leadership, visioning and planning • Deployment: ERP package selection, communication, process management, training and education, project management, legacy systems management, system integration, system testing, cultural and structural changes • Evaluation: performance evaluation and management

Table 11.1. (continued)

Study	Methodology	Critical Success Factors Identified
Brown and Vessey (2003)	Case study (3 organisations)	*Identified five factors for successful ERP implementations:* • Top management is engaged in the project, not just involved • Project leaders are veterans, and team members are decision makers • Third parties fill gaps in expertise and transfer their knowledge • Change management goes hand-in-hand with project planning • A satisfying mindset prevails
Umble et al. (2003)	Case Study (1 organisation)	*Identified critical factors for successful ERP implementations* • Clear understanding of strategic goals, commitment by top management, excellent project management, organisational change management, a great implementation team, data accuracy, extensive education and training, focused performance measures, multi-site issues
Somers and Nelson (2004)	Survey (116 organisations)	*Identified and tested the relative importance of the key players and activities across the ERP project life cycle, which affect the success of these projects.* • Key players: top management, project champion, steering committee, implementation consultants, project team, vendor-customer partnerships, vendors' tool, and vendor support • Key activities: user training and education, management of expectations, careful selection of the appropriate package, project management, customisation, data analysis and conversion, business process re-engineering, defining architecture, dedicating resources, change management, establishing clear goals and objectives, education on new business processes, interdepartmental communication and cooperation

Table 11.1. (continued)

Study	Methodology	Critical Success Factors Identified
Gargeya and Brady (2005)	Content analysis (Secondary data pertaining to SAP implementations in 44 companies)	*Identified six common factors that are indicative of successful or non-successful SAP implementations:* • Lack of appropriate culture and organisational readiness is the most important factor contributing to failure of SAP implementations • The presence of project management approaches and appropriate culture and organisational readiness are the most important factors contributing to the success of SAP implementations
Motwani et al. (2005)	Case study (1 organisation)	*Identified the factors that facilitated the success of ERP implementations and examined the factors that initially inhibited the success of the implementation process and explained how these barriers were overcome.* • Strategic initiatives, learning capacity, cultural readiness, IT leveragability and knowledge sharing capacity, network relationships, change management practices, process management practices
Tsai et al. (2005)	Survey (multiple organisations)	*Identified the critical factors causing failure in the implementation of the enterprise resource planning (ERP) system. Suggested that companies should focus on improving the management of these failure factors to increase the rate of success in the implementation of the ERP systems.* • Time frame and project management • Personnel training • Change management

Table 11.2. Major Studies Examining the Role of Culture in ERP Implementations

Study	Methodology	Findings
Soh et al. (2000)	Case study (1 organisation)	*Discussed the cultural misfits of ERP packages from a Singaporean perspective. Identified the:* • Different types of misfits employed: data, process, and output • Resolution strategies employed • Impacts on organisations

Huang and Palvia (2001)	Literature Review and Theoretical Framework Development	*Identified a range of issues concerning ERP implementations by making a comparison of advanced and developing countries.*
		• National and environmental factors:
		  o Current economic status and economic growth
		  o Infrastructure
		  o Government regulations
		• Organisational and internal factors
		  o Low IT maturity
		  o Small firm size
		  o Lack of process management and BPR experience

Davison (2002)	Case study (1 organisation)	*Compared educational ERP system implementation practices in North America and Hong Kong. Identified certain differences along the following dimensions:* • Access to information • Homonyms (meanings associated with numbers) • Re-engineering and empowerment
Martinsons (2004)	Case study (8 organisations) Review of the results of a survey (189 organisations)	*Investigated the ERP implementations in China, comparing the practices of state-owned enterprises and private enterprises.* • There is a poor fit between ERP systems and traditional Chinese management systems. • Identified 8 differences between state-owned and private enterprises in terms of: Primary project aims, role of top management, role of steering committees, role of consultants, scope of implementation, pace of implementation, implementation problems, and evaluation and outcomes.
O'Kane and Roeber (2004)	Case study and Survey (1 organisation)	*Focused on an ERP implementation in a Korean company* • Determined what impact natural culture has on the implementation process of ERP systems by testing some of the propositions developed by Davison (2002) and Martinsons (2004).

Table 11.2. (continued)

Study	Methodology	Findings
Reimers (2003)	Survey (80 organisations)	Investigated the crucial implementation process and context variables which warrant closer study of ERP enabled organisational change in China. • Ownership is strongly associated with implementation process characteristics • Project governance (role and decision making style of steering committee) affects implementation success
Liang et al. (2004)	Interviews (5 organisations)	*Investigated the five companies that attempted to implement foreign ERP systems with unsuccessful results. Identified the following types of problems for failure:* • Language problems • Report format and content problems • Cost control module problems • Price problems • Business process redesign problems • Customer support problems • Consulting partner problems
Rajapakse and Seddon (2005)	Case study (6 organisations)	*Utilizing Hofstede's dimensions of national culture, investigated the impact of national and organisational culture on the adoption of western-based ERP software in developing countries in Asia.* • The findings revealed a clash of cultural forces between the culture embedded in western products and the culture of Asian ERP adopters. • Four pairs of opposing cultural forces work against ERP implementations in Asia: ○ Centralized vs. decentralized ○ Low vs. high level of accountability and discipline ○ Low vs. high level of commitment ○ Low vs. high level of change

11.6 References

Allen D, Kern T, Havenhand M, (2002) ERP Critical Success Factors: An exploration of the Contextual Factors in Public Sector Institutions. Proceedings of the 35th Hawaii International Conference on System Sciences

Al-Mashari M, Al-Mudimigh A, Zairi M, (2003) Enterprise resource planning: A taxonomy of critical factors. European Journal of Operational Research 146(2):352–364

Amin, N, Hinton M, Hall P, Newton M, Kayae R, (1999) A Study of Strategic and Decision-Making Issues in Adoption of ERP Systems Resulting from a Merger in the Financial Services Sector. 1st International Workshop on Enterprise Management Resource and Planning Systems (EMRPS), Venice, Italy:173–181

Bingi P, Sharma M, Godla J, (1999) Critical Issues Affecting an ERP Implementation. Information Systems Management 16:3:7–8

Brown C, Vessey I, (1999) ERP Implementation Approaches: Toward a Contingency Framework. Proceedings of the International Conference on Information Systems:411–416

Brown C, Vessey I, (2003) Managing the Next Wave of Enterprise Systems: Leveraging Lessons from ERP. MIS Quarterly Executive 2(1):65–77

Davison R, (2002) Cultural Complications of ERP. Communications of the ACM 45:7:109–111

De Mooij M, (2000) The future is predictable for international marketers: Converging incomes lead to diverging consumer behavior. International Marketing Review 17 (2):103–113

Dubé L, Paré G, (2003) Rigor in Information Systems Positivist Case Research: Current Practices, Trends, and Recommendations. MIS Quarterly 27(4):597–635

Eisenhardt KM (1989) Building Theories from Case Study Research. The Academy of Management Review 14(4):532–550

Esteves J, Pastor J, (2000) Towards unification of critical success factors for ERP implementations. Proceedings of the 10th Annual Business Information Technology (BIT) Conference, Manchester, UK:44–52

Esteves J, Pastor J, (2001) Enterprise resource-planning systems research: an annotated bibliography. Communications of the AIS 78:1–52

Gargeya VB, Brady C, (2005) Success and failure factors of adopting SAP in ERP system implementation. Business Process Management Journal 11(5):501–516

Hofstede G, (1991) Culture and organisations: Software of the mind. London, UK: McGraw Hill

Hofstede G, (2001) Culture's Consequences: Comparing Values, Behaviors, Institutions, and Organisations across Nations. 2nd Ed., Sage Publications, London, England

Holland C, Light B, (1999) Critical Success Factors Model for ERP Implementation. IEEE Software May/June:1630–1636

Huang Z, Palvia P, (2001) ERP implementation issues in advanced and developing countries. Business Process Management Journal 7(3):276–84

Klaus H, Rosemann M, Gable GG, (2000) What is ERP?. Information Systems Frontiers 2(2):141–162

Lee A, (2000) Researchable Directions for ERP and Other New Information Technologies. MIS Quarterly 24(1):3–8

Liang H, Xue Y, Boulton WR, Byrd TA, (2004) Why Western vendors don't dominate China's ERP market?. Communications of the ACM 47(7):69–72

Maitland C, Bauer J, (2001) National level culture and global diffusion: The case of the Internet. In Charles Ess (Ed.), Culture, technology, communication: Towards an intercultural global villagew. Albany, NY: State University of New York Press:87–128

Markus ML, Tannis C, (2000) The Enterprise Systems Experience – From Adoption to Success. In Framing the Domains of IT Research: Glimpsing the Future through the Past, R. W. Zmud (Ed.), Cincinnati, OH: Pinnaflex Educational Resources, Inc.

Martinsons MG, (2004) ERP in China: One Package, Two Profiles. Communications of the ACM 47(7):65–68

Molla A, Loukis I, (2005). Success and Failure of ERP Technology Transfer: A Framework for Analyzing Congruence of Host and System Culture. Development Informatics Working Paper Series

Motwani J, Akbul AY, Nidumolu V, (2005) Successful implementation of ERP systems: a case study of an international automotive manufacturer. International Journal of Automotive Technology and Management 5(4):375–386

Murray MG, Coffin GWA, (2001) Case Study Analysis of Factors for Success in ERP System Implementations. Proceedings of the Americas Conference on Information Systems, August 3-5, Boston, Massachusetts:1012–1018

O'Kane JF, Roeber M, (2004) ERP Implementations and cultural influences: a case study. 2nd world conference on POM, Cancun, Mexico

Parr A, Shanks G, (2000) A Model of ERP Project Implementation. Journal of Information Technology 15:289–303

Rajapakse J, Seddon PB, (2005) ERP Adoption in Developing Countries in Asia: A Cultural Misfit. Available at http://gebennehmen.de/PLAYOUGH/ERP_vs_Culture.pdf, accessed 12/10/07

Reimers K, (2003) International Examples of Large-Scale Systems – Theory and Practice I: Implementing ERP Systems in China. Communications of the AIS 11(20):335–356

Robey D, Ross J, Boudreau M, (2002) Learning to Implement Enterprise Systems: An Exploratory Study of the Dialectics of Change. Journal of Management Information Systems 19(1):17–46

Roseman M, Sedera W, Gable G, (2001) Critical Success Factors of Process Modeling for Enterprise Systems. Proceedings of the Americas Conference on Information Systems, August 3-5, Boston, Massachusetts:1128–1130

Sankar CS, Raju PK, Nair A, Patton D, Bleidung N, (2005) Enterprise Information Systems and Engineering Design at Briggs & Stratton: K11 Engine Development. JITCAR, 7(1):21–38

Sharma R, Palvia P, Salam AF, (2002) ERP Selection at Custom Fabrics. JITCA 4(2):45–59

Soh C, Kien SS, Tay-Yap J, (2000) Enterprise Resource Planning: Cultural Fits and Misfits: Is ERP a Universal Solution?. Communications of the ACM 43(4):47–51

Somers TM, Nelson KG (2004) A taxonomy of players and activities across the ERP project life cycle. Information and Management 41:257–278

Sumner M, (2004) Enterprise Resource Planning. Pearson, Prentice Hall, Upper Saddle River, New Jersey

Tatsiopoulos I, Panayiotou N, Kirytopoulos K, Tsitsiriggos K, (2003) Risk Management as a Strategic Issue for the Implementation of ERP Systems: A Case Study from the Oil Industry. International Journal of Risk Assessment and Management 4(1):20–35

Tsai W, Chien S. Hsu P, Leu J, (2005) Identification of critical failure factors in the implementation of enterprise resource planning (ERP) system in Taiwan's industries. International Journal of Management and Enterprise Development 2(2):219–239

Umble E, Haft R, Umble M, (2003) Enterprise Resource Planning: Implementation Procedures and Critical Success Factors. European Journal of Operational Research 146(2):241–257

Veiga JF, Floyd S, Dechant K, (2001) Towards modelling the effects of national culture on IT implementation and acceptance. Journal of Information Technology 16(3):145–158

Xue Y, Liang H, Boulton WR, Snyder CA, (2005) ERP Implementation Failures in China: Case studies with Implications for ERP Vendors. International Journal of Production Economics 97(3):279–295

Yin R, (2003) Case Study Research: Design and Methods. Sage Publications, California

Appendix

Utilisation of Suchman's Paper

Séverine Le Loarne[1], Audrey Becuwe[2]
[1]Grenoble Ecole de Management
[2]Ecole des Dirigeants et Créateurs d'Entreprise (EDC Paris)

Table A.1 Utilisation of Suchman's paper

Authors, date and review of publication	Object of article	Mobilisation and quotation of Suchman's paper	Thesis of article and results
Pourder R., John CHS, (1996), *Academy of Management Review*	Develop an evolutionary model that contrasts hot spot and non-hot spot competitors within the same industry.	*"As the emerging industry sub-population gains legitimacy within the region, access to capital and market improves".*	Initially, economies of agglomeration, institutional forces, and managers' mental models create an innovative environment within the hot spot. Over time, those same forces create a homogeneous macroculture that suppresses innovation, making hot spot competitors more susceptible that non-hot spot competitors to environment jolts.

Table A.1 (continued)

Authors, date and review of publication	Object of article	Mobilisation and quotation of Suchman's paper	Thesis of article and results
Brown AD, (1997), *Academy of Management Review*	Theory of narcissism employed to analyse the dynamics of group and organisational behaviour.	*"the idea that organisation must exhibit 'congruence' or 'isomorphism' with the social values and norms of acceptable behaviour in the larger social system is well established"* The use of Suchman's article is very generic. It is not related to the topic of legitimacy but on how organisations adapt themselves to social norms and values.	Organisational identification permits organisational legitimisation.
Reed R, Lemak DJ, Hesser WA, (1997), *Academy of Management Review*	Shift in mission of the U.S. nuclear weapons complex from the production of nuclear materials and weapons to one of environmental cleanups.	*"legitimacy rests on a foundation of satisfying the self-interests of the organisation's audiences, having a positive evaluation of the organisation and its activities, and receiving positive backing"*.	Draw attention to the management and social issues the complex is facing in the related areas of organisation-culture change, the public's health fears and the management of risks.
Sahay S, Walsham G, (1997), *Organization Studies*	Social structure and managerial agency in India.		Describe possible influences that social structure has on the shaping of managerial attitudes in India. This framework is then used to provide the lens through which a specific Indian-government-initiated, information-technology project is analysed.

Table A.1 (continued)

Authors, date and review of publication	Object of article	Mobilisation and quotation of Suchman's paper	Thesis of article and results
Ruef M, Scott WR, (1998), *Administrative Science Quaterly*	*Organisational legitimacy*: the antecedents and effects of two forms of organisational legitimacy: managerial and technical.	Three quotations: -*"these and related contributions represent considerable diversity but also reflect a common underlying conception, which has been formulated by Suchman as follows: "legitimacy is a generalised perception or assumption that the actions of an entity are desirable, proper, or appropriate within some socially constructed system of norms, values, beliefs and definitions".* -*"As Suchman noted, legitimacy is a "generalised perception" representing the "reactions of observers to the organisation as they see it, thus, legitimacy is possessed objectively, yet created subjectively"* -*"As Oliver (1991) and Suchman (1995) have proposed and Elsbach and Sutton (1992) have demonstrated, organisations are not simply passive recipients in legitimisation processes but work actively to influence and manipulate the normative assessments they receive from their multiple audiences"* So, Ruef and Scott use the legitimacy definition of Suchman.	The antecedents of legitimacy vary, depending on the nature of the institutional environment as well as the organisational function that is being legitimated.

Table A.1 (continued)

Authors, date and review of publication	Object of article	Mobilisation and quotation of Suchman's paper	Thesis of article and results
Barron DN, (1998), *Organization Studies*	Processes by which two organisational forms in New York City can become legitimate: credit unions and the Morris Plan Bank.	- *"Second, although ecologists have tended to see legitimacy as cognitive, other scholars have considered it to imply moral or pragmatic acceptance of an organisational form (Suchman 1995)".* - *"Following Suchman (1995), I call these three forms pragmatic, moral, and cognitive legitimacy, respectively. Pragmatic legitimacy 'rests on the self-interested calculations of an organisation's most immediate audiences' (Suchman, 1995: 578)".* - *"Potential members, customers, or sponsors of an organisation must believe that such an involvement will be in their interests. Moral legitimacy 'reflects a positive normative evaluation of the organisation and its activities' (Suchman, 1995: 579). This is perhaps the definition of legitimacy that is closest to its meaning in common usage".*	Various mechanisms affected different types of legitimacy
Mone MA, McKinley W, Barker III VR, (1998), *Academy of Management Review*	Organisational decline	*"institutional theorists point out that organisations are subject to institutionalised expectations about what behaviours they can pursue legitimately"*	Develop a contingency model which identifies variables as the environmental, organisational and individual levels of analysis that determine whether organisational decline inhibits or stimulate innovation.

Table A.1 (continued)

Authors, date and review of publication	Object of article	Mobilisation and quotation of Suchman's paper	Thesis of article and results
Beckert J, (1999), *Organization Studies*	Question of how to deal with interest-driven behaviour and institutional change	*"Even if entrepreneurs reflect upon the constraining qualities of institutionalised practices, and "management of legitimacy" (Suchman, 1995) has to take into account the negative consequences resulting from violations of institutionalised demands".*	Develop an integrative concept which theorises the connection of strategic agency and institutions in a model of institutional change.
Kostova T, Zaheer S, (1999), *Academy of Management Review*	*Organisational legitimacy*	*" traditionally, researchers have examined legitimacy at two levels: (1) at the level of classes of organisations and, (2) at the organisational level (Suchman, 1995)"* .	Obtaining legitimacy is both a socio-political and cognitive process through which the environment and the organisation continually test and redefine the legitimisation process. The organisation is involved in a continual process of interpreting and influencing its own actions as they are related to the legitimising requirements of the larger environment.

Table A.1 (continued)

Authors, date and review of publication	Object of article	Mobilisation and quotation of Suchman's paper	Thesis of article and results
Scott SG, Lane VR, (2000), *Academy of Management Review*	Analyse organisational identity from the perspective of manager-stakeholder relationships.	" *similar to legitimacy, organisational identity is objectively held – that is, it has a reality independent of individual observers – although it is subjectively arrived at.* "	Develop a model of organisational identity construction that reframes organisational identity within the broader context of manager-stakeholder relationships, and which draws attention to organisational identity as negotiated cognitive images and to embeddedness of organisational identity within different systems of organisational membership and meaning.
McKinley W, Zhao J, Garrett Rust K, (2000), *Academy of Management Review*	Understand the phenomenon of organisational downsising	Two quotations: -" *with some exceptions, there has been little effort to model the specific cognitive processes that underlie convergence toward taken-for-grantedness in managerial practices or organisational forms*" -"*In some cases the expectations attain enough cognitive legitimacy that alternatives to the expected practice are literally unthinkable*"	With their socio-cognitive model, they argue that downsizing has become institutionalised through the collectivisation and reification of a "downsizing is effective" schema.

Table A.1 (continued)

Authors, date and review of publication	Object of article	Mobilisation and quotation of Suchman's paper	Thesis of article and results
Mazza C, Alvares JL, (2000), *Organization Studies*	Explore the role of popular press in the production and legitimisation of management ideas and practices.	*"Legitimacy and legitimisation appear, therefore, as toolboxes where any researcher can find the definition that better fits his or her purposes* (Suchman, 1995; Massa, 1998)"	They argue that popular press is the arena where the legitimacy of management ideas and practices is produced.
Hasselbladh H, Kalinikos J, (2000), *Organization Studies*	Critical approaches various neo-institutional accounts of the process of formal organising		Develop a framework that seeks to outline the conceptual means for decomposing the carriers of rationalised patterns, models and techniques and showing the distinctive ways in which they implicate the building blocks of formal organising.

Table A.1 (continued)

Authors, date and review of publication	Object of article	Mobilisation and quotation of Suchman's paper	Thesis of article and results
Schneiberg M, Bartley T, (2001), *The American Journal of Sociology*	Assess three approaches to state regulation: capture theory, interest group analyses, and neoinstitutional research.	*" For neoinstitutionalists, legitimacy – the alignment of sectors with prevailing principles of rational or just order and the positive evaluation, credibility, or certification, which stem from that alignment – is what drives the adoption of structures or policies"*	Develop a theory of how political and institutional conditions shape industries' governance options. They analyse state policy and economic order as a result of multilevel political organisation, and the activation of controversy, legitimacy crises, and anti-company forces within institutional fields – rather than as an expression or reflection of taken-for-granted understandings. Such an approach supports a more political and contested view of institutional factors, highlighting how political and institutional processes fundamentally define and transform the choice sets available for private and public problem-solving behaviour.

Table A.1 (continued)

Authors, date and review of publication	Object of article	Mobilisation and quotation of Suchman's paper	Thesis of article and results
Lawrence TB, Win MI, Devereaux Jennings P, (2001), *Academy of Management Review*	Examine the relationship between time and processes of institutionalisation.	" *While Luke emphasises the ability of elites to manipulate those under them, other researchers have demonstrated the potential for a wide variety of organisational participants to manage meaning (Suchman, 1995) through language and culture*".	They argue that pace and stability, two temporal dimensions of institutionalisation, depend on the mechanism used by agents to support the institutionalisation support.
Jones C, (2001), *Organization Studies*	Gain a better understanding of the co-evolutionary processes of entrepreneurial careers, institutional rules and competitive dynamics in emerging industries.		A co-evolutionary perspective was integrated with insights from institutional and resource-based theories to explain how the American film industry emerged, set an initial trajectory with specific institutional rules and competitive dynamics, and then changed.
Walgenbach P, (2001), *Organization Studies*	Study on the use of ISO 9000 standards and ISO 9000 certification in Germany		The implementation of the ISO 9000 standards was regarded as an occasion for structuring and led to the development of a system of bureaucratic control that was both enabling and coercive.

Table A.1 (continued)

Authors, date and review of publication	Object of article	Mobilisation and quotation of Suchman's paper	Thesis of article and results
Hensmans M., (2003), *Organization Studies*		*"Unobtrusive strategy in this sense, means framing participant's interest in such a fashion so as to make them feel what is being claimed is not underly conflictive, but, on the contrary, credible, appropriate, comprehensive and desirable"* (Lukes, 74; Fliegstein, 97; Rao, 1998 and Suchman, 1995)	
Zajac E.J. and Westphal J.D. (2004), *American Sociological Review*	A social constructionist view of financial market behaviour. In particular, they seek to extend neoinstitutional theory in two ways: (1) link the social dynamics of financial markets with the processes and outcomes of institutionalisation, and (2) show how the phenomenon of institutional decoupling is related to the process of institutionalisation.	*"thus, subsequent policies that appear to conform to the same logic enjoy greater social acceptance, and firms realise greater legitimacy benefits from adopting them"*	Their study posits that institutionalisation processes might increase the market value of a policy as more firms adopt it, despite growing evidence of decoupling. They propose that investors are likely to reference prior market reactions to similar events in estimating the reactions of other investors to the adoption of the focal policy, and they further propose that this social estimation process causes the value of corporate policies to become increasingly taken-for-granted, even as the rate of decoupling increases over time.

Table A.1 (continued)

Authors, date and review of publication	Object of article	Mobilisation and quotation of Suchman's paper	Thesis of article and results
Phillips N, Lawrence TB, Hardy C, (2004), *Academy of Management Review*	discourse analysis and the process of institutionalisation.	*"As Suchman (1995) argues, the management of legitimacy depends on communication as actors instrumentally deploy evocative symbols to garner legitimacy".*	They argue that language is fundamental to institutionalisation. They develop a discursive model of institutionalisation that highlights the relationships among texts, discourses, institutions and actions.

Index

Printed in the United States
124440LV00003B/244-267/P

9 781848 001824